ASTRONOMY WITH A HOME TELESCOPE

ASTRONOMY
WITH A HOME TELESCOPE

THE TOP 50 CELESTIAL BODIES
TO DISCOVER IN THE NIGHT SKY

SETH PENRICKE

ZEPHYROS
PRESS

CONTENTS

ORION NEBULA

INTRODUCTION

Cliffs erode and shorelines change, but the stars look the same now as they did 10,000 years ago. When you look up at the night sky, you're seeing the same constellations that Shakespeare, Cleopatra, and Julius Caesar saw.

Of course, our understanding of the cosmos has expanded greatly since the time of Cleopatra, who never knew what made stars burn or that the Earth revolves around the Sun. But it may surprise you to know that those discoveries, and many other great astronomical findings, were made with only a very weak telescope or, in many cases, no telescope at all.

Gazing into a telescope only as powerful as a pair of small binoculars, Galileo discovered the orbiting moons of Jupiter, which convinced him that the **heliocentric** (Sun-centered) model of our solar system was correct. And simply by observing the planets and how they drifted about the sky—without a telescope at all—Johannes Kepler developed a theory that they orbit in ellipses, not circles. His theory of orbits, postulated in 1609, remains true today.

Since the dawn of civilization, humans have looked to the stars out of curiosity and for wisdom and guidance. Before television, electricity, and the printing press, the night sky was the main form of evening entertainment, and early humans spent so much time looking at it that they developed all kinds of ideas about its meaning. The planets were named for Greek and Roman gods. The seven-day week was based on the seven objects in the night sky that move independently of each other—the Sun, the Moon, Mars, Mercury, Jupiter, Venus, and Saturn—in contrast to a fixed field of stars.

ASTRONOMY COMPELS THE SOUL TO LOOK UPWARD AND LEADS US FROM THIS WORLD TO ANOTHER.

— PLATO

To see the original greatest show on Earth, you need only look up. Even the unaided eye grants you access to a world above, infinite and endlessly exciting, that billions of your ancestors looked at and pondered with awe.

With just a $20 pair of binoculars you can see the crescent phases of Venus as it revolves around the Sun, and the oval shape made by Saturn's rings. Also, one of the brightest stars in the sky, Sirius—also known as the Dog Star—will reveal itself to be a binary, two suns locked in an eternal tango.

With the lowest magnification telescope you can buy, you'll see the moons of Jupiter and Saturn as they spin around their respective planets. And Saturn's rings will become as clear to you as they were to Dutch astronomer Christiaan Huygens in 1659, when he became the first person to recognize them.

With a telescope of a slightly higher power, you'll be able to see the big storm on Jupiter—also known as Jupiter's Great Red Spot. First observed by German amateur astronomer Samuel Heinrich Schwabe in 1831, it's likely this storm has been raging since long before humans gained the ability to see it. Using this level of magnification, many objects in the sky that look like single points of light will expand into binaries, clusters, and even whole galaxies.

With higher resolutions and magnifications, you'll be amazed at what you can see: asteroids, comets, nebula, even planets eclipsing nearby stars. Many comet hunters and planet hunters are amateur astronomers, including Thomas Bopp, codiscoverer of one of the brightest comets of the past century, the comet Hale-Bopp.

As you will see, the universe is a bustling and intriguing place, full of wonder and information. Many of the most important astronomical discoveries, including that of Pluto, Jupiter's red spot, and nearby asteroids, were made using simple telescopes placed on lawns and roofs by people like you. Indeed, it is impressive how much of the universe you can discover from the comfort of your own backyard.

SEEING
STARS

PREPARE TO BE DAZZLED

Astronomy with a Home Telescope will introduce you to some of the most interesting sights in the universe that are visible from Earth's **Northern Hemisphere** (the half of the Earth above the equator). The 50 celestial objects highlighted here were chosen because they are bright, accessible, impressive, and diverse. So whether you live in a bright city or a small town, whether you prefer long observations that change over time or quick views now and then, you'll find just the thing in the pages that follow.

Think of the list of celestial objects in this book as a "greatest hits" collection of sights found in the night sky. To keep it interesting, the chosen celestial bodies vary greatly in size and quality—from the largest objects in the universe (galaxies) to the smallest visible objects in the solar system (comets). All should be visible with a small telescope, and many of them can be seen with the naked eye. However, some of the objects and occurrences—especially the Sun and eclipses—should not be viewed directly and require special filters when observing through a telescope.

Individual stargazing experiences can vary depending on a multitude of factors, including your eyesight, your location on Earth, the elevation, the time of year, the amount of cloud coverage, your equipment's telescopic power and stability, and, most important, the level of light pollution. As already stated, this guide is geared for observations from the Northern Hemisphere, where 90 percent of Earth's population lives. It should be noted, though, that the sky from the Southern Hemisphere abounds with its own celestial wonders, such as the violent center of the Milky Way and the two satellite galaxies that orbit ours—the Large and Small Magellanic Clouds. (For a list of resources about Southern Hemisphere stargazing, see page 166).

Are you ready to expand your horizons and discover the wonders of the night sky? Let's get started!

YOUR TRUSTY TELESCOPE

Your fascinating and inspiring tour of the universe starts with a basic understanding of your telescope. The compass is to the ship captain what the telescope is to the astronomer—a navigational instrument and a source of entertainment all in one.

Don't worry, you don't need to know much about lenses or optics to stargaze, but you will find that the more you do know, the faster you'll be able to find stars and calibrate a telescope.

Setup

As already mentioned, a good pair of binoculars will help you see a large part of our solar system, with the added benefit of their portability and **stereoscopic** (two-eyed) view. However, on a spinning planet like Earth, it can prove difficult to keep them steady long enough to observe small and fainter objects. So, while binoculars are well suited to spotting larger objects—like the ridges and craters of the Moon, or the many moons of Jupiter—they are not as great for viewing faint astronomical bodies. Since a larger lens size is typically required to see fainter objects, you'll have to invest in a telescope to accomplish that.

Telescopes come in two basic types: **refracting** and **reflecting**. Refractors use **lenses** (small, clear panes of glass) to magnify the sky; reflectors magnify light through curved mirrors. Most observatories use reflecting telescopes. Some telescopes, known as **catadioptric telescopes**, use a combination of lenses and mirrors. While the difference is not too important for the beginner, knowing what kind of telescope you use can increase your ability to find stars quickly and maintain your equipment.

Many telescopes today, even relatively inexpensive ones, are **computerized telescopes** with digital equipment that identifies stars and automatically positions the telescope. This way, finding a celestial body becomes as easy as entering a few coordinates. Some can even tell you what object you are looking at after you point at it, making identification extremely simple. The coordinate systems that these computerized telescopes might use is discussed more on page 16.

Regardless of the type of telescope you have, the crucial first step is making sure it is mounted on a flat surface. Nearly all telescopes will rest on a tripod; if the telescope is not level, adjust the legs of the tripod until it is. Some tripods and telescopes have built-in levels to help you position them correctly. Likewise, levels are built into many smartphones, which have gyroscopes inside.

If you're viewing from your backyard, it may be easier to set up your telescope while it is light outside. Note that if you set up during the daytime, you should leave the lens cap on your telescope or, at least, keep it pointing as far from the Sun as possible. Telescopes can be damaged if pointed at the Sun for even a brief moment. If your telescope has a digital camera inside, Sun exposure could destroy it perminantly.

Once your telescope is level, make sure it is securely fastened to the tripod mount, usually with a custom screw. Some telescopes can be very fragile and may not recover from a fall.

Viewing

Before you even look through your scope, it's important to let your eyes adjust to the dark. The human eye has two different regulators that control your ability to see in the dark: The first is the pupil, which reacts within a few seconds to changes in light. The secondary darkness adaptation occurs in the retina of the eye and takes about 20 minutes to fully adjust, which means that you will need to be patient, turn off your computer or phone, and take some time to enjoy the night before stargazing.

Another way to maximize your viewing potential is to learn when to look away. This sounds counterintuitive, but directing your gaze slightly away from an object will make the object appear brighter. Interestingly, the focal point of the human eye tends to be less sensitive than the periphery.

SCOUTING FOR YOUR VIEWING LOCATION

In general, a dark, Moonless night or one with a new Moon are better times for stargazing (unless your intent is to look at the Moon). Regions with less **light pollution** (unwanted light inhibiting visibility) are better locations for stargazing, as are higher elevations where there is less atmospheric disturbance.

GETTING YOUR BEARINGS IN THE UNIVERSE

In general, most objects that orbit our Sun will appear in the sky along the plane of the **ecliptic**, the wide line across the sky through which all bodies that orbit the Sun travel. Because all planets in our solar system travel along roughly the same plane, they all cut a similar path running along a band of sky about 8 degrees in width, or about the distance between your pinky and thumb held at arm's length.

Since the Sun, Moon, and planets all follow this plane, as long as you can identify at least the Sun or Moon, you should have a general directional idea of where the eliptic sits in the sky. As the seasons change, the ecliptic moves slightly up or down due to the Earth's tilt, but no matter where you are on Earth, it will always be above you, since it runs in a ring around the planet. Any amateur astronomer should have an understanding of where and what the ecliptic is—and knowing this can also save your life if you are lost, since it points to where the Sun and other celestial bodies rise (east) and set (west).

The **zodiac**, the 12 constellations symbolically important to astrology, appear in the background of the ecliptic, meaning the planets, the Sun, and the Moon move past them throughout the year.

For millennia, human civilizations have creatively imagined all visible stars to be part of different constellations. Hence, the quick and easy way to find any star is to locate it in relation to its nearby constellations. In this book, directions for locating celestial bodies are given based on the constellations they belong to and the time of year when you are stargazing.

LIGHT POLLUTION

Whether hobbyist or professional, astronomers have to consider the effect of light pollution on their observations. Light pollution can obscure visibility and render faint stars almost invisible, depending on cloud coverage and the phase of the Moon. When stargazing from your own house or backyard, you can minimize light pollution by turning off indoor and outdoor light sources, especially lights that shine up rather than down toward Earth.

Beyond your home, also consider how local streetlights affect viewing, as streetlights are a major source of light pollution. Certain cities in proximity to scientific observatories—including San Jose, California, and Tucson, Arizona—have passed ordinances to limit light pollution from streetlights, usually by mandating downward-pointing sulfur streetlamps that emit low-intensity light. You can write to your local government to suggest this change, if they haven't done so already. Sulfur lighting is also the cheapest in terms of energy costs.

If you live in a hopelessly light-polluted area, such as a large, sprawling metropolis, you may find that occasionally you'd like to travel somewhere else to observe the night sky. As a general rule, the higher the elevation and the farther from a metropolis, the better, keeping in mind that cloud coverage is also a major factor. Elevation is important because light pollution is compounded as it travels through more atmosphere, and atmospheric distortion goes down as elevation rises. This is why the best observatories are placed on tall mountains, far from population centers—as in Mauna Kea in Hawaii, or Kitt Peak in southern Arizona.

USING COORDINATES TO FIND CELESTIAL BODIES

Many telescopes, especially digital ones, use fundamental measurement systems that do not reference constellations as previously described. Following is a brief overview that may be helpful for more advanced astronomy learners or those with a computerized telescope—many of which allow you to enter coordinates directly or even look up celestial objects by their common names.

Astronomers have several quantitative systems for locating celestial objects. The oldest one is the **equatorial coordinate system**. If you are familiar with latitude and longitude as a system for measuring points on Earth, the equatorial coordinate system is similar—except it maps points in space above the Earth, using two measurements: **declination** and **right ascension** (often abbreviated D and RA).

- Declination, an analog to latitude, is measured above and below the equator and varies from -90 degrees to 90 degrees. If you're in the Northern Hemisphere, you will probably not see objects far below 0 degrees.

- Right ascension, the analog to longitude, uses hours instead of degrees, since one revolution of the Earth constitutes 24 hours. Think of it this way: In any 24-hour period the star that appears directly above you at hour 0 should be there again at hour 24.

The other system for locating celestial objects is the **horizontal coordinate system**, which makes reference to the positions of the horizon and the Earth. The horizontal coordinate system is expressed in terms of **altitude** and **azimuth**.

- The altitude, measured in degrees between 0 and 90, is the relative position of the object above the horizon.

- The azimuth, measured in degrees between 0 and 360, expresses how far east from due north the object is located.

If you are given coordinates to a celestial body, it should be fairly clear which coordinate system is used. If it contains hours, it is equatorial; if it contains two sets of degrees, it is horizontal.

Some digital telescopes take hours to set up and match to coordinates, while constellations can be found and referenced in less than a minute. That's why this book uses constellations as reference points instead of coordinates. If you have a digital telescope that takes input coordinates, you can use a simple Internet search to find them for each of the celestial bodies in this book.

START EXPLORING

Once your telescope is set up and your eyes have adjusted to the dark (yes, those 20 minutes can seem like an eternity), you are ready to start exploring. For most telescopes, the process of finding a target and sighting it is as follows:

1. In the "directions" section of a celestial body entry (pages 28 to 153), find the constellation that references the celestial body. For example, Polaris (North Star) (see page 47) references the Little Dipper in its "directions" section. This will be your reference point.

2. With the naked eye, locate this constellation in the sky. Find the reference star, if the target is not a visible star.

3. Point your telescope's viewfinder at the target. Look up with your unaided eye and then move the telescope over the target you've spotted. It is generally much easier to start with the naked eye and then activate the telescope, rather than to get lost looking at an enhanced sky through the telescope first.

4. If your telescope is on a mount or tripod, tighten the mounting knobs enough so it can still move on the mount but does not tilt down when you let go. Keep your hand on the knob, and when you find your target, tighten it slightly.

5. If your telescope has a focus, adjust it now until the view is clear. Most telescopes with a focus have a single knob that adjusts the focal length, making the field of view either more or less fuzzy. If you can't get the telescope to focus on a distant target, it might help to start with an object on Earth, like a tree or mountain. If the Moon is visible, that can also work well. Adjust for this object until it is

in focus. Then move to the stars behind it and adjust until they are clear. When things look clear, go back to step 2. If you can't get the view to look clear, your telescope may need some cleaning or maintenance; refer to the Cleaning and Maintenance section below for advice.

6. Do you see what you expected? If yes, tighten the mount knob. If not, go back to step 2. Remember that when you look just slightly away, the target is often enhanced in your peripheral vision.

7. Be aware that because of Earth's rotation, after a few minutes, the target may move out of your field of view, depending on its size. So, over time you may have to recalibrate. Some computerized telescopes automatically adjust for the movement of Earth.

Cleaning and Maintenance

CLEANING THE LENS

To see the skies clearly, it is crucial to maintain a clean, dust-free telescope lens. As with eyeglass lenses or laptop screens, the best way to clean a telescope lens or viewfinder is with lens cleaning fluid and a microfiber cloth. Microfiber cloths are especially important since they do not scratch lenses. Even a seemingly soft material, like paper towel, can scratch a lens if used with force—and even a tiny scratch will be evident on a good telescope. Spray the lens with the cleaning fluid and then move the microfiber cloth in circles as you wipe off the lens, radiating outward from the center. If the microfiber cloth gets very dirty, you can wash it in the washing machine and hang it to dry.

MAINTENANCE AND PARTS REPLACEMENT

Because of the wide variety of telescope designs, it is hard to write a comprehensive guide to maintenance. However, no matter what type you are using, here are two universal indicators that your telescope is in need of repair or replacement:

1. **AN INABILITY TO FOCUS ON ANY OBJECT.** This may point to an alignment issue or broken knob. Depending on the telescope model and root cause of the problem, you will need to either replace your telescope or have it professionally repaired.

2. **A SCRATCH ON A LENS OR MIRROR.** This can be a fatal blow to a telescope. Usually this requires a replacement mirror or lens, which can be much cheaper than buying a new telescope. A scratch will be very obvious in a telescopic lens, particularly when looking at large objects that take up the whole field of view (such as the Moon). Check with the manufacturer about getting a replacement lens. These can vary in cost from $40 to $200 (for large-lens telescopes).

Computerized telescopes have more complicated maintenance needs. Certain telescopes need power to function, which usually comes from either a battery or an outlet. Likewise, some very advanced computerized telescopes need an Internet or GPS connection in order to calibrate and adjust with the sky.

Accessories

Filters are the most common telescope accessories. Because objects in the sky have variable levels of light, certain sights require a filter in order to view them more accurately. For instance, the Moon is a very bright object, so bright that in order for you to view it comfortably through a large telescope, you need a darker filter. Polarizing filters are also helpful for viewing planets, which tend to be quite bright in contrast to other objects in the sky. They act like dimmer switches to lower an object's brightness to a comfortable level.

Viewing the Sun (see page 144) and solar eclipses (see page 148) directly can be very dangerous to the eyes. Therefore, when directing a telescope at these objects and events, it is extremely important that you use a special **solar filter**. A solar filter blocks almost all light and allows you to see granulation on the Sun, the solar corona, and all other sorts of strange and wonderful solar phenomena.

With the advent of digital optics, many new telescope models feature digital imaging capabilities. With these telescopes you can take snapshots of what you see or download data from the Internet on newly found or fast-moving objects like comets and natural satellites.

You can also modify many analog telescopes by attaching a digital camera or connecting it to a computer. If you are interested in contributing to the asteroid-spotting or comet-spotting community, these accessories are very helpful.

OTHER STARGAZING OPTIONS

As you know, many celestial bodies are visible to the naked eye. Your ability to see is mediated only by how long you let your eyes adjust to the dark: 15 to 20 minutes of adjustment should reveal much of the dark night sky.

When mounted on a tripod, a digital camera with a good lens and high resolution can reveal many smaller stars and the dust bands of the Milky Way with long-exposure or time-lapse shots. Once these images are processed, you can zoom in on various regions on your computer, and stargaze digitally. This can often be helpful for picking out dim or hard-to-see objects—the challenge then becomes identifying the objects on the screen, rather than in the sky.

As mentioned earlier, a pair of binoculars is the best way to stargaze on a low budget. First, spot your celestial object on your own, then bring the binoculars to your eyes. Binoculars don't have large lenses like telescopes, but even through their modest lenses you can see a number of planets' moons that are invisible to the naked eye, including Jupiter's.

For each of the 50 celestial objects in this book, you'll find directions for viewing with the naked eye, binoculars, or a low- or high-powered telescope. Except for a few particularly elusive objects, almost every one is visible to different degrees by any of these means.

ASTRONOMY BY PHONE

Whether by convenient design or cosmic coincidence, many smartphones have cameras that fit perfectly into the eyepiece of your telescope. Since even inexpensive modern smartphones have truly phenomenal digital cameras, you can often capture what you see in your eyepiece with the camera on your phone. Of course, there are much more expensive attachments and telescopes with built-in digital cameras, but for astronomy on a budget, a smartphone can be an indispensable resource.

The only caveat is that the brightness of a smartphone can bring your eyes out of their dark-adjusted state. No worries—there's a trick to getting around that. The human eye's pupil does not constrict in response to red light the same way it does in response to white and blue light. If you cover your smartphone screen with red plastic wrap, you can still use it to snap photographs without losing your ability to see dim stars.

If your smartphone has a time-lapse feature, or if it has enough memory to record a movie, you may be able to capture fast-moving astronomical objects, like the inner moons of Jupiter or the slowly changing shadows of the Moon.

2

THE TOP 50 CELESTIAL BODIES

WHAT'S OUT THERE?

What do you think about when you look to the night sky? There are as many answers to this as there are observers. It is a place of stunning natural beauty and wonder—and endless possibilities for imagination. Your journey through the heavens, as guided by this book, will teach you, fascinate you, and hopefully leave you wanting more. This is just the beginning of your adventure with the sky.

Using This Book

The objects in this book are presented in order of their difficulty to spot. First is the Moon—one of the easiest to see—then other objects in our solar system, and then readily visible stars that serve as reference points for more difficult objects you can't always see with the naked eye, such as star clusters, nebulae, and, finally, other galaxies.

Many of the following celestial bodies can be seen with the naked eye. In fact, because of the nature of light pollution, you may find it easier to see some of these objects unaided in a dark location than with a telescope in a light-polluted area (like New York City or Los Angeles). Variations on what you might see are noted in the description for each entry.

Additionally, most of the stars, nebulae, and galaxies are relatively static in appearance, while planets and moons vary depending on the night. These variations are noted, too, in each entry's description.

To inform your observation, accompanying each celestial object entry are some qualifying factors:

VISIBILITY The ease with which one can see the object, with notes on viewing in different light conditions or with different telescope models. Visibility is rated on a scale of ● ○ ○ ○ ○ to ● ● ● ● ●, with ● ● ● ● ● being the easiest to see.

SEASON The time(s) of year in the Northern Hemisphere when the celestial object is most visible. While stars have reliable seasons in which they are visible, objects within our solar system have different calendars unrelated to the tilt of the Earth's axis. This is described more in the entries on specific planets.

QUALITY Notes on brightness, color, intensity, and variation of intensity for **variable stars**.

DISTANCE FROM EARTH For distant objects, this is measured in **light-years**—the distance that light will travel in one year in a vacuum (approximately 9.5 trillion kilometers or 5 trillion miles). For objects in our own solar system this distance is measured in kilometers and miles.

DIRECTIONS Information that will help you spot the object in reference to major stars and constellations. For additional help locating objects in the night sky, refer to your favorite astronomy web site (such as www.astroviewer.com) or stargazing app (such as Star Chart or Night Sky Lite).

SIGNIFICANCE The celestial body's historical, cultural, and scientific importance.

ORIGIN STORY Linguistic roots of the body's name and any mythological connections.

POP CULTURE Current references to the celestial object that are fun and informative.

[
REMEMBER:
Check the season and visibility notes before you begin,
as certain winter stars are not visible in summer and
vice versa. Let's discover what's out there!
]

THE TOP 50

#	NAME	TYPE	PAGE
1	Moon	Moon	28
2	Jupiter & its moons	Planet	31
3	Venus	Planet	34
4	Mars	Planet	36
5	Saturn & its moons	Planet	38
6	Mercury	Planet	41
7	Uranus	Planet	43
8	Neptune	Planet	45
9	Polaris	Star	47
10	Alpha Centauri	Star	51
11	Sirius A & B	Star	53
12	Mizar & Alcor	Star	55
13	Fomalhaut	Star	58
14	Betelgeuse	Star	62
15	Rigel	Star	64
16	Albireo	Star	66

#	NAME	TYPE	PAGE
17	Pollux & Castor	Star	68
18	Vega	Star	72
19	Aldebaran	Star	74
20	Arcturus	Star	76
21	Capella	Star	78
22	Algol	Star	80
23	Mira	Star	82
24	Perseus Double Cluster	Cluster	86
25	Hercules Globular Cluster	Cluster	88
26	Dumbbell Nebula	Nebula	91
27	Beehive Cluster	Cluster	93
28	Hyades Cluster	Cluster	95
29	Iris Nebula	Nebula	97
30	Omega Nebula	Nebula	100
31	Wild Duck Cluster	Cluster	102
32	Trifid Nebula	Nebula	105
33	M4	Cluster	108

#	NAME	TYPE	PAGE
34	Pleiades	Cluster	111
35	Orion Nebula	Nebula	113
36	Ghost of Jupiter	Nebula	115
37	Ring Nebula	Nebula	117
38	Eagle Nebula	Nebula	119
39	Lagoon Nebula	Nebula	121
40	Crab Nebula	Nebula	124
41	Owl Nebula	Nebula	126
42	Virgo Cluster	Cluster	128
43	Bode's Galaxy	Galaxy	131
44	Andromeda Galaxy	Galaxy	135
45	Triangulum Galaxy	Galaxy	137
46	Whirlpool Galaxy	Galaxy	139
47	Comets	Comet	141
48	The Sun	Star	144
49	Solar Eclipse	Phenomenon	148
50	Lunar Eclipse	Phenomenon	152

1

THE MOON

VISIBILITY ● ● ● ● ●

SEASON All.

QUALITY Off-white and gray in color, and occasionally orange-red when rising or setting in areas with high air pollution. Becomes dark red during a lunar eclipse (see page 152).

DISTANCE FROM EARTH 370,000 kilometers (230,000 miles).

DIRECTIONS You can determine when the Moon will rise and set based on which phase it is in. Full Moons tend to rise on an opposite schedule to the Sun—meaning at dusk—whereas new Moons occur during the day since the Sun must be behind the Moon (from Earth's perspective) in order to be unlit. The Moon follows the ecliptic, along with the Sun and all the planets.

SIGNIFICANCE The Moon is perhaps the most studied astronomical object in history. It is visible day and night and from anywhere on Earth. It is, along with the Sun, one of the two most visible objects in the night sky. The length of the calendar month is based around the Moon's cycle, which averages 28 days. The Hebrew calendar is still based on the Moon's phases.

The Moon is also the only celestial body (so far) that human astronauts have landed on and explored—six times between 1969 and 1972. While the tracks and machinery left behind by astronauts are not visible with a telescope, you can view the regions they explored.

The Moon is the first celestial object explored in this book because it has an effect on the visibility of other objects in the sky—on a Moonlit night, the light pollution from the Moon affects what you can see elsewhere. A very bright Moon limits your ability to see dim stars. If you know when the Moon rises and sets, take this into account when planning your night sky viewing.

The Moon is in tidal lock with the Earth, meaning it rotates only once for each time it revolves around Earth. This means you only ever see one side of the Moon—the "light side," as it is sometimes called. The "dark side" of the Moon, which does also receive light from the Sun, is never visible from the perspective of Earth.

The darker parts of the Moon are called *mare*, Latin for *sea*, since at the time they were first discovered they were thought to be oceans. We now know they are simply valleys of dark volcanic rock. Looking up with your telescope, you should be able to pick out the brightest ones easily: Oceanus Procellarum (Ocean of Storms) and Mare Imbrium (Sea of Showers). These surround three large craters: Kepler, Aristarchus, and Copernicus. Tycho is another large crater, which appears on the south edge of the Moon and looks like a white-gray circle with white lines emanating from it.

Much of the Moon's topography is visible with binoculars. However, viewing the Moon in its crescent phases with a powerful telescope can be more rewarding, as you can make out the shadows from the mountains and valleys. If you observe the same location during different phases, you should be able to see the location of the shadows change, further highlighting the topography.

ORIGIN STORY It is hard to assign an exact date to the Moon's origin. However, the word *moon* dates back several thousand years in European languages. It is also a general term for the satellite of any planet. The Moon is frequently associated with femininity and also madness—the term *lunacy* is related to the word *lunar*, as that trait was thought to be affected in intensity based on the phases of the Moon.

POP CULTURE In the twentieth century, Arthur C. Clarke's *2001: A Space Odyssey* imagined a black monolith on the Moon, placed there by an alien civilization. Rock band Pink Floyd's album *Dark Side of the Moon* (1973), one of the best-selling albums of all time, played on the symbolism of the unseen half. In folklore, a full Moon causes werewolves to transform. Monday is named for the Moon.

2

JUPITER
& ITS MOONS

VISIBILITY ● ● ● ● ●

SEASON Early spring to summer until 2019.

QUALITY Bluish when viewed through a weak telescope, faintly reddish-white when viewed through a strong telescope, with four points of light surrounding it (its largest moons).

DISTANCE FROM EARTH Between 600 million kilometers (370 million miles) and 930 million kilometers (580 million miles) depending on its location relative to Earth as it revolves around the Sun.

DIRECTIONS Jupiter follows the ecliptic like all planets. It moves in a 12-year cycle, meaning it passes approximately one zodiacal constellation each year. Its moons cling closely to it and should usually be visible, unless they are blocked by the planet. From 2015 until 2018, Jupiter will move from Cancer (2015) toward Leo (2016), Virgo (2017), and then Libra (2018).

SIGNIFICANCE Jupiter is the king of planets, being both larger than all other planets in mass and second brightest in the sky (after Venus, whose brightness varies much more than Jupiter). While composed mostly of hydrogen and helium, Jupiter is thought to have a rocky core about the size of Earth. The discovery of its moons, easily visible with a pair of binoculars, led early astronomer Galileo to conclude that not all celestial bodies orbited the Earth—at the time, a radical idea.

A strong telescope allows you to see the Great Red Spot, Jupiter's most prominent feature. The Great Red Spot is a giant storm that has been circulating for centuries—as long as the spot has been visible to human astronomers. Recently it has started to shrink slightly for reasons not fully understood.

Jupiter's surface is covered by wispy gases that whip smoothly around the planet. If you can take pictures with your telescope, repeated viewings might reveal changing surface features.

Jupiter has 67 moons and counting. The four largest moons, named the Galilean moons, for their discoverer, are the only ones readily visible from Earth. Together, these moons—Callisto, Ganymede, Europa, and Io—are some of the largest rocky bodies in our solar system; in fact, Ganymede is larger in volume than the planet Mercury. Amateur astronomers can figure out which moon is which by tracking their paths from night to night. Io, the nearest moon, has an orbital period of 1.8 days; Europa, the second-closest, has an orbital period of 3.6 days; Ganymede has an orbital period of 7.2 days; and Callisto has an orbital period of 16.7 days. If you observe a bright point of light on one side of Jupiter one evening, and it appears on the other side the next night, this object is almost certainly Io.

The Galilean moons of Jupiter have been cited as ideal places for human colonies. Despite being far from the Sun, these moons retain heat through regular expansion and contraction as an effect of Jupiter's gravity.

ORIGIN STORY Jupiter was the primary god in Roman mythology and in Greek mythology, too, though the Greeks called the planet Zeus. In Modern Greek, the planet is still called Zeus.

POP CULTURE Jupiter's moons have as many pop culture references as the planet does. In the film *2001: A Space Odyssey*, the crew of the *Discovery One*, including HAL 9000, set off to observe a monolith in orbit around Jupiter. In the *Star Trek* series, Jupiter is home to a prominent space station. In Romance languages, Thursday is named after Jupiter.

3

VENUS

VISIBILITY ● ● ● ● ●

SEASON Varies, but when visible, will generally follow the Sun by a few hours or precede it by the same length.

QUALITY Often the brightest object in the sky, and yellowish upon closer inspection. Its crescent phases, like those of the Moon, are readily visible with only a pair of binoculars.

DISTANCE FROM EARTH 40 million kilometers (28 million miles) to 250 million kilometers (160 million miles).

DIRECTIONS Venus can be found along the ecliptic. Because it is closer to the Sun than Earth is, it will never be seen rising more than a half day before the Sun or setting more than a half day after.

SIGNIFICANCE Though Mars gets much of the attention these days from Earth's space programs, Venus is not without its own charms. Venus is a sister planet to Earth. With nearly the same size and mass, it also has roughly the same gravity and is rich in carbon dioxide, much like Earth in its youth. Venus is thought, perhaps, to have had liquid water in its early years. And like Earth, it has a liquid core and a surface topography shaped by volcanoes.

However, the Venusian atmosphere leaves much to be desired—its pressure is astoundingly high and choked with thick, noxious sulfur dioxide gas. Temperatures range from 750 degrees Fahrenheit (400 degrees Celsius) and above, higher than the melting point of lead. Venus's dense toxic clouds create a barrage of lightning strikes that pound the surface constantly—Soviet probe *Venera 12* even heard thunderclaps as it approached the surface.

Because of the dense atmosphere and winds that surround it, the surface of Venus cannot be observed directly with a telescope. However, its cloudy upper atmosphere glows a yellowish-white. Venus is an interesting observational target because its phases are so vivid from Earth—the planet often appears as a beautiful arced crescent, and repeated observations will reveal its changing phases. Venus even occasionally eclipses the Sun from Earth's standpoint—these eclipses can be seen only with a solar filter (see page 144, The Sun). During these eclipses, Venus will appear as a black dot drifting in front of the Sun.

ORIGIN STORY Venus was the Roman goddess of fertility, beauty, and love. The temple built in her honor in the second century AD is the largest ancient Roman temple in recorded history.

POP CULTURE In contemporary Western culture, Venus is associated with femininity, in contrast to the masculine Mars. For example, the best-selling book about relationships between men and women is titled *Men Are from Mars, Women Are from Venus* (1992). Many consumer goods marketed toward women include Venus in their names—including razors and lipsticks. The symbol of the planet, a circle with a cross extending downward, is also the biological symbol for female. "Friday" is derived from the word for Venus in many European languages.

4

MARS

VISIBILITY ● ● ● ● ●

SEASON Varies depending on the year but always moves along the ecliptic.

QUALITY Mars is easy to spot because of its distinct orange-red color and its situation along the ecliptic.

DISTANCE FROM EARTH 60 million kilometers (35 million miles) to 390 million kilometers (240 million miles).

DIRECTIONS Mars appears along the ecliptic and moves relatively quickly along it, traversing several zodiac constellations per month. Check online or in a daily newspaper for more exact information. Most newspapers contain night sky information on visible planets, usually near the weather section.

SIGNIFICANCE Mars, known as the Red Planet due to its reddish, iron-rich crust, has been the subject of the most intense scientific scrutiny of any other planet in the solar system. Two active vehicles rove the surface, the NASA landers *Curiosity* and *Opportunity*, while five satellite probes orbit above.

The second-closest planet to Earth, Mars is considered the next logical step, after the Moon, for a manned mission. With a day similar in length to Earth's (25 hours) and both water and carbon dioxide present on its surface, Mars is a tempting site for explorers—provided that they can withstand the howling wind storms, chilly evenings, and thin atmosphere. Still, its tourist sites are numerous and include the tallest mountain in our solar system, Olympus Mons, visible with a strong telescope as a faint coloration near the planet's equator.

Of particular interest to amateur astronomers are Mars's polar regions. They are snow white and easily visible due to their highly reflective nature. The region just below the white north pole is a desert and appears as a slightly darker ring. Unlike Venus's visible phases, Mars will always appear as full or mostly full, given that it is farther from the Sun than Earth. Because it varies much in its distance to Earth, Mars is easiest to view when it is in **opposition** to Earth—when the Sun, Earth, and Mars form a line, with the Sun and Mars at either end and Earth in the middle.

Mars has two small moons, likely captured asteroids, named Phobos and Deimos. Both orbit very closely to their home planet and are difficult to view, except at certain angles and in dark conditions.

ORIGIN STORY Mars was the Roman god of war, known as Ares to the ancient Greeks.

POP CULTURE Mars's symbol is a circle with an arrow pointing from the top, identical to the symbol for male in certain Western cultures. A candy bar is named for the planet, as is the month of March. Mars frequently serves as the source of fictional alien invaders, perhaps most famously in Orson Welles's radio broadcast *The War of the Worlds*, which was mistakenly believed to be real news by many listeners at the time of its debut. Tuesday was named for Mars, based on the Latin *Martis*.

5

SATURN

& ITS MOONS

VISIBILITY ● ● ● ● ●

SEASON Fall and winter.

QUALITY Yellowish, with visible rings through a decent telescope.

DISTANCE FROM EARTH About 1.5 billion kilometers
(930 million miles).

DIRECTIONS As with any planet orbiting the Sun, Saturn moves slowly
through the plane of the ecliptic, passing through constellations as it
goes. From 2015 to 2023, it will move from the vicinity of Libra, through
Sagittarius (2018), and finally to the vicinity of Capricorn (through
2023). Saturn orbits the Sun once every 30 years, which means that it
will reappear in the same region of the sky every 30 years.

SIGNIFICANCE Saturn's beautiful and strange ring system has fascinated astronomers for centuries. Though not fully understood, it
is theorized that the rings have existed for billions of years—perhaps

since the dawn of our solar system. Composed mainly of billions of individual particles of water and ice and varying in thickness between a few centimeters and 1 kilometer, the rings measure more than 70,000 kilometers (44,000 miles) from side to side—about six times that of Earth's diameter. Interestingly, however, the mass of Saturn's rings is very small overall, equal to about 2 percent of the mass of all the water on Earth.

Titan, Saturn's largest moon, is larger in volume than the planet Mercury, and it has the densest atmosphere of any moon in our solar system. Aside from our own Moon, it is the only moon on which a man-made probe has landed: In 2005, the European Space Agency's *Huygens* probe took pictures from the foggy surface.

One of the joys of viewing Saturn through a telescope is discovering its nonspherical shape. Saturn is more oblate than most planets, meaning it is wider than it is tall. When viewed in profile it appears oval. With a weak telescope, it may look like three circles, or a long oval-shaped blob; this is how Galileo saw it, not recognizing its rings. With a slightly better lens—a 3-inch telescope should be enough—the rings should be clearly visible. With a telescope lens that is 7 inches or longer, you may be able to make out the black gap in the middle of its rings. This gap is known as the Cassini Division.

Titan is readily visible with a 2-inch telescope lens and appears as a point of light trailing Saturn. It orbits Saturn in 16 days, and repeated viewings will give you a glimpse of its orbit in motion. The largest lenses will reveal Saturn's six largest moons (out of 62), in order of mass from greatest to least: Titan, Rhea, Iapetus, Dione, Tethys, and Enceladus.

ORIGIN STORY Saturn was named for the Roman god of liberation, wealth, and agriculture, Cronus in Greek mythology. The Saturnalia festival was a grand celebration in his honor.

POP CULTURE The English word *Saturday* is derived from Saturn. The movie *Interstellar* (2014) imagined a wormhole orbiting near Saturn. Kurt Vonnegut's 1959 novel *The Sirens of Titan* was partially set on the moon Titan.

LAGOON NEBULA

6

MERCURY

VISIBILITY ● ● ● ◉ ◉

SEASON Varies; Mercury closely follows the Sun and will always be close to it in the sky.

QUALITY White in color, but due to its proximity to the horizon, it is often orangish. Appears in different phases, like Venus.

DISTANCE FROM EARTH 100 million kilometers (60 million miles) to 220 million kilometers (135 million miles).

DIRECTIONS To find Mercury, follow the Sun and look along the ecliptic at sunset or before sunrise. Mercury tends to be most visible when it is at its greatest **elongation**—when it forms a triangle with the Sun and Earth, rather than a straight line.

SIGNIFICANCE The closest planet to the Sun and also the smallest, Mercury is a hot, rocky world with almost no atmosphere. Littered with craters, it resembles the Moon in color and reflection from close up. Though it is the closest planet to the Sun, it is not uniformly hot. The unlit side of the planet and the valleys at the poles remain well below freezing, and a permanent sheath of ice covers part of the northern pole.

Because it is so close to the Sun, this planet is hard to reach via spacecraft. Only two probes have visited—*MESSENGER* and *Mariner* 10, neither of which still orbit the planet (*MESSENGER* crash-landed in 2015). The twin satellite probes *Mercury Planet Orbiter* and *Mercury Magnetospheric Orbiter* will arrive in 2017.

As an inner planet, Mercury will appear to have phases when viewed from Earth—like the Moon, it waxes and wanes in its crescent. It is slightly harder to view than the other planets, but not by much—it simply needs to be timed carefully. Through a telescope, Mercury will appear as a crescent in the dusk or dawn sky.

Mercury is an excellent candidate for viewing during a total solar eclipse (see page 148). Unless it is behind the Sun, it will become visible during totality.

[
WARNING:
NEVER point your telescope anywhere near Mercury
when the Sun is still up. Accidentally looking at the Sun,
even for a second, can permanently blind you.
Wait until the Sun has set to view Mercury.
]

ORIGIN STORY Named after the Roman god Mercury, Hermes in Greek mythology, who was the messenger of the gods. Several East Asian cultures knew it as the "water star."

POP CULTURE In many languages across the world, the word for *Wednesday* is associated with Mercury. In astrology, Mercury in **retrograde**—when it appears to move from west to east rather than the usual east to west—is associated with upset plans and problems. The phrase *Mercury must be in retrograde* has entered the pop culture lexicon as a convenient excuse for when something goes wrong.

7

URANUS

VISIBILITY ● ● ● ● ●

SEASON Fall and winter.

QUALITY Not visible with the naked eye, except in the darkest regions of Earth, but visible with a small telescope or even binoculars, if you know where to look. Appears white or bluish-green, depending on Earth's atmosphere.

DISTANCE FROM EARTH About 3 billion kilometers (1.9 billion miles).

DIRECTIONS Since Uranus has such a long orbital period, its place in the stars remains relatively fixed—it appears in Pisces until 2020.

SIGNIFICANCE The seventh planet from the Sun and the first planet invisible to the naked eye to be discovered, Uranus is a cold, distant gas giant, with a rocky core similar in size and composition to that of Earth's. It has a cool, serene, blue glow, as opposed to the yellow-orange palettes of Jupiter and Saturn—partly as a result of the ammonia and methane components of its atmosphere. Though it appears placid, looks can be deceiving. Winds whip the outer atmosphere at velocities approaching the speed of sound—around 500 miles per hour!

Outer planets like Uranus are more challenging to see. Nonetheless, Uranus is actually visible with binoculars if you know where to look. Look along the ecliptic to Pisces, then northwest of the star Alrisha (at the corner of Pisces). Uranus will be a blue-green disk in the vicinity of Pisces until around 2020, when it will appear closer to the constellation Aries, where it will remain for many years.

Uranus has four satellites that should be visible with a strong home telescope, listed here in order of increasing difficulty to spot: Titania, Oberon, Ariel, and Umbriel. Though faint, they will appear as points of light in line with Uranus's plane of orbit and the ecliptic.

ORIGIN STORY Uranus is named after the Greek god of the sky. All 27 of its moons are named after characters in William Shakespeare and Alexander Pope's plays: For instance, Titania and Oberon are fairies in Shakespeare's *A Midsummer Night's Dream*, and Ariel and Umbriel are sylphs in Pope's *The Rape of the Lock*.

POP CULTURE The radioactive element uranium is named after Uranus. Titania is featured in Kim Stanley Robinson's book *Blue Mars*, and Uranus and its moons have featured in multiple *Doctor Who* storylines over the show's many seasons.

NEPTUNE

VISIBILITY ● ● ● ● ●

SEASON Autumn, for the next few decades.

QUALITY Faint, but very blue.

DISTANCE FROM EARTH About 4.5 billion kilometers (2.8 billion miles).

DIRECTIONS Look in the vicinity of Aquarius, where it will be until the 2020s, when it will begin to move closer to Pisces. It cuts a line along the ecliptic south of the Aquarian star Ancha and the north of Skat.

SIGNIFICANCE Neptune, the farthest planet from the sun in our solar system, since Pluto's demotion to dwarf-planet status in 2006, is quite similar to Uranus in composition. Its atmosphere, composed of methane, ammonia, helium, and hydrogen, surrounds a rocky core about the size of Earth. However, Neptune's upper atmosphere suffers from large storms that create dark spots about the diameter of Earth on its surface, which resemble bruises. These "Great Dark Spots," as they are known, are thought to circulate, disappear, and reappear with some regularity. The winds that whip Neptune's surface are the fastest and strongest in our solar system, gusting up to 1,300 miles per hour! Neptune has a faint ring system that is not visible with a telescope.

Neptune is the most distant body in our solar system that is still visible with binoculars, but only if you know exactly where to look. Neptune appears along the ecliptic and currently in the vicinity of Aquarius, where it will remain for a few years until it migrates closer to Pisces. It should always appear as a full disk, bluish in color. Stronger telescopes should reveal variations in surface coloration and, occasionally, the dark spot.

Triton, Neptune's largest satellite by far, is also the most peculiar in our entire solar system. It is the largest object, moon, or planet that orbits in the "wrong" direction—that is, the opposite orbital direction of the Sun and Neptune's other 13 moons. Triton is also one of few moons that is geologically active and may have volcanoes on its surface. It should be visible on a dark night with a large reflector (8 inches or larger). Triton appears as a small gray dot hovering near the planet.

ORIGIN STORY Neptune was the Romanized version of the Greek god Poseidon, ruler of the sea. The planet was formally discovered in 1846 by French and English astronomers and was predicted to exist based on observations of wobbles in Uranus's orbit that hinted at the nearby gravitational pull of another body. Interestingly, Galileo actually observed Neptune near Jupiter in 1613 but did not realize what it was at the time. Triton is named after the Greek messenger god and son of Poseidon, who carried a trident. All other Neptunian satellites are similarly named after oceanic deities.

POP CULTURE Neptune and its purple Neptunian inhabitants played a prominent role in the twenty-first-century science fiction comedy cartoon series *Futurama*. Samuel R. Delany's 1976 novel *Trouble on Triton* may be the best-known fictional depiction of this moon.

9

POLARIS

THE NORTH STAR

VISIBILITY ● ● ● ● ●

SEASON Always visible at night regardless of season, except in the summer in northernmost latitudes where the Sun never sets.

QUALITY White or yellowish-white.

DISTANCE FROM EARTH Approximately 400 light-years.

DIRECTIONS If you can, find the Little Dipper constellation (Ursa Minor): The North Star is its tail—the last point of light on the "handle." If you can't find the Little Dipper, look for the Big Dipper (Ursa Major): The side of its ladle forms part of a line that points toward Polaris.

SIGNIFICANCE Though not the brightest star in the sky, Polaris may be the most important star in the Northern Hemisphere, simply because of its navigational significance. Throughout the course of the day the sky shifts as Earth rotates, but the location of Polaris remains precisely the same. If you were to take a long exposure of the night sky, Polaris would be a fixed point, while the other stars would all appear as blurred lines.

This occurs not because of any special property of the star, but simply because of its unique situation directly above Earth's northern pole. Stars situated directly above a planet's poles are known as **pole stars**. As Earth's orbit wobbles and our solar system moves through the Milky Way, Polaris will cease to be a true northern pole star—lucky for us, this event is still hundreds of years away. Polaris has been the North Star for about two millennia. Approximately 5,000 years ago, Thuban was the North Star. In about 1,000 years, Gamma Cephei will become the North Star, followed by Iota Cephei (in the fifth millennium AD), and, eventually, Vega.

Note that there is no navigational equivalent in the Southern Hemisphere, though the Southern Cross comes close and has a similar navigational usefulness.

Polaris is also unique for being a type of variable star known as a **Cepheid variable**. This means that its brightness changes slightly, with a predictable regularity. It is the closest of its kind in proximity to Earth. In the case of Polaris, this brightness shift happens because of a cycle of cooled and heated gas that affects its luminosity. Repeated careful observations may allow you to measure and determine the period of variability with some accuracy.

Polaris is actually a multiple star, with several dimmer companions. The second-largest, Polaris B, can be seen in dark areas with the naked eye but reveals itself more clearly with a telescope. It is about one-fourth of the mass of Polaris A and closer to our Sun in mass. Polaris B orbits Polaris A at quite a distance—about 360 billion kilometers (220 billion miles), or 2,400 times as far as Earth's distance from the Sun. Polaris Ab, the very closely orbiting **white dwarf** companion to Polaris A, will probably not be visible with your home telescope. Ditto for the dim Polaris C and Polaris D.

ORIGIN STORY Polaris has been known since ancient times and was catalogued by Ptolemy in the second century AD. Its name is Latin and means "near the pole."

POP CULTURE Polaris appears in the *Green Lantern* comics as home to the Blue Lantern Corps. A science fiction convention named for the star is held annually in Toronto, Canada. Polaris is also the name of a comic book character in the Marvel universe and line of telescopes manufactured by Meade.

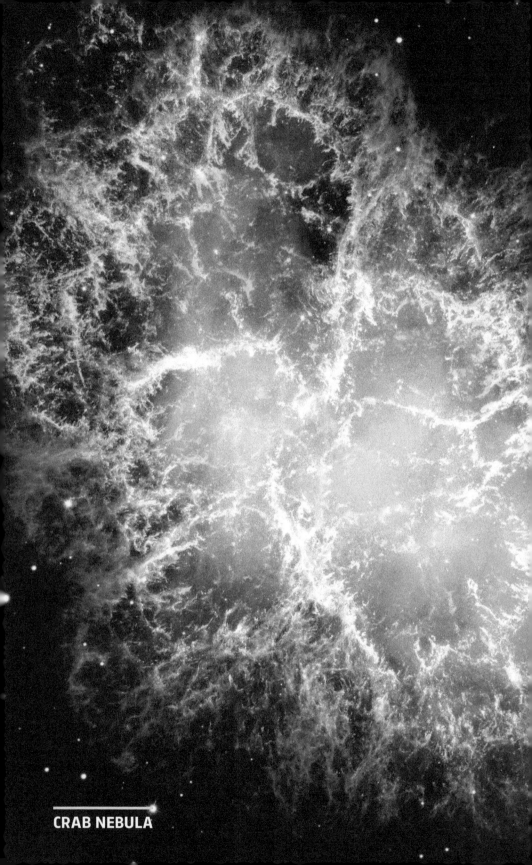
CRAB NEBULA

10

ALPHA CENTAURI

VISIBILITY ● ● ● ● ● Visible only from very southern locations in the Northern Hemisphere, such as Hawaii, southern Florida, Mexico, and Puerto Rico.

SEASON Spring (May especially).

QUALITY In the rare instances in which it is visible, Alpha Centauri is one of the brightest stars in the sky and will appear close to the horizon. Its color is yellow to yellowish-white.

DISTANCE FROM EARTH 4.37 light-years.

DIRECTIONS The brightest star in the Centaurus constellation, Alpha Centauri is visible in southern latitudes in North America. Look to Centaurus in the season in which it is visible.

SIGNIFICANCE Though difficult to see from the Northern Hemisphere, Alpha Centauri is worthy of inclusion if only for its fame as a potential human destination—as the closest star to Earth, it is also subject to the most scrutiny and speculation as to the nature of its solar system. It is a binary star, made up of Alpha Centauri A and Alpha Centauri B. The pair has an orbital period of approximately 80 years, and they are separated by about the same distance as the Sun and Saturn. It may also have one planet attached, Alpha Centauri Bb, though its existence is not confirmed.

Alpha Centauri is most visible in fall and winter in the Southern Hemisphere. Because it sits so low in the sky from the perspective of those in the Northern Hemisphere, there is a brief window during the late spring in which it can be seen from the southern latitudes of the United States. This peaks in May, when it should appear in the constellation Centaurus at approximately 1 a.m. (time may vary depending on your exact location). Look for it just above the horizon.

With a telescope, you can easily see the binary nature of this star. Alpha Centauri A is yellow and similar to our Sun in size. Alpha Centauri B is smaller than the Sun and slightly cooler and dimmer.

A third, even dimmer companion, the **red dwarf** Proxima Centauri, is suspected to be an orbital companion to Alpha Centauri. If this is true, Proxima Centauri's orbital period would be near 500,000 years. It is visible in longer exposures as a red dot but is very hard to differentiate from background stars.

ORIGIN STORY The centaur, a half-horse, half-man creature from Greek mythology, is the constellation's namesake. As the brightest star in the Centaurus constellation, Alpha Centauri is appended with the first letter of the Greek alphabet.

POP CULTURE A large US naval ship is named for the Centaurus constellation. A popular 1990s video game, *Alpha Centauri*, revolved around the premise that humans would start a second colony on an orbiting planet of Alpha Centauri. Similarly, in the popular *Civilization* series of strategy games, one route of victory was to colonize Alpha Centauri. The characters of the TV series *Lost in Space* were en route to Alpha Centauri when they were sent off course. The *Avatar* film series takes place on the moon of a fictional planet in Alpha Centauri.

11

SIRIUS A & B

VISIBILITY ● ● ● ● ●

SEASON Winter and spring.

QUALITY White and twinkling.

DISTANCE FROM EARTH 8.6 light-years.

DIRECTIONS Sirius is the eye of the dog in the constellation Canis Majoris. Orion's belt appears to point backward toward Sirius.

SIGNIFICANCE As the brightest star in the night sky, Sirius—also known as the Dog Star—is hard to miss. Within the constellation of Canis Majoris, Sirius is one of the few stars that most people can name, partly due to its historical significance. The Ancient Egyptian calendar is based around the time of year that it rose in the sky, and the Greeks and Romans held festivals around the star, which they considered a bad omen. The phrase *dog days* originated from the belief that Sirius, in certain positions in the sky, adversely affects the behavior of dogs.

In the nineteenth century, improved telescope technology revealed Sirius to be a binary star. This had been suspected based on observations of Sirius wobbling slightly in its orbit, an indication that it was being pulled by another object's gravity. Its companion star is a white dwarf labeled Sirius B that is as massive as the Sun but as small and compact as Earth. As a result, Sirius B is incredibly dense and hot. Sirius B—also known as the "pup"—is difficult to see with a home telescope by virtue of its small size. A 1-foot aperture or greater can discern it relatively easily as a small dot just off to the side of Sirius A.

In winter and spring Sirius appears in the evening. In summer it appears early, just before dawn. Because of its incredible brightness it often appears to twinkle or radiate different colors, an effect of distortions from Earth's atmosphere. In fact, some ancient accounts of Sirius refer to it as "red," though this is suspected to be a result of distortion and the way its white light seems to glow all colors if you stare at it for a while.

Sirius is one of few stars visible in the daytime and can be spotted if you know precisely where to look. If viewing with a telescope in the daytime, point it far away from the Sun. Because it is so far south, daytime viewing is easier in the Southern Hemisphere.

ORIGIN STORY The word *sirius* means "scorching" in Ancient Greek.

POP CULTURE A satellite radio company is named after this star. It is also a common name for dogs. In Homer's *Iliad*, Sirius is said to portend the Trojan War. Sirius Black, godfather to title character in the Harry Potter series of books and films, is named after the star.

12

MIZAR & ALCOR

THE HORSE & RIDER

VISIBILITY ● ● ● ● ●

SEASON Year-round, but most visible in spring and winter.

QUALITY Two whitish-blue stars.

DISTANCE FROM EARTH 83 light-years.

DIRECTIONS The second star in the handle of the Big Dipper, Mizar, and its slightly dimmer companion Alcor, should be discernible as two points of light very close to each other.

SIGNIFICANCE Mizar and Alcor are an odd pair in more ways than one. Though they appear as a binary, even to the naked eye, they are in fact a configuration of six stars. The Big Dipper, and therefore Mizar and

Alcor, are visible for most of the year, though the time that it rises varies depending on the season. The pair is separated by about one light-year, and they are gravitationally bound to each other, meaning they slowly orbit a common point.

When you've located Mizar, see if you can discern Alcor, the dimmer star, with the naked eye. It's about one-quarter of a degree away from Mizar. On a dark night, if you have good vision, this shouldn't be too difficult to spot. Next, point your lens in their direction. What do you see? Discerning Mizar and Alcor as separate stars is a great first test for focusing a telescope.

Once you differentiate Alcor from Mizar, the next step is to view Mizar as two separate stars. Mizar is a visible binary, but it can't be seen with the naked eye. With a weak telescope, Mizar easily separates into Mizar A and Mizar B. This binary was one of the first ever discovered, and Galileo himself observed and commented on it.

Now you should see three stars where once there appeared to be only one. But what about the other three? Mizar and Alcor are actually a sextuple system. And, indeed, Mizar A and Mizar B are both binaries themselves. However, it is very difficult to **split** either of these two stars into their respective binaries with a ground telescope. As for Alcor's binary companion, it's nearly impossible: It was only in 2009, and with the most powerful telescope in the world, that Alcor B, a tiny, dim red dwarf, was discovered.

Mizar and Alcor are well-known stars in astronomy, and many cultures around the world have ascribed different meanings to them. Interestingly, several cultures saw the ability to discern this binary with the naked eye as a sort of test. In Japanese culture, Alcor is called the Lifespan Star, and it is thought that someone who cannot see it will die within the year. In the Middle Ages, a Persian astronomer wrote of Alcor as a star used to test eyesight.

ORIGIN STORY *Mizar* comes from the Arabic word for waist. *Alcor* has a complicated etymological history but may be related to the Arabic *al khawwar*, which means "faint one." The name *Horse and Rider* seems to date to antiquity, however several cultures have legends relating to the stars. A German legend tells of a wagon driver named Hans (associated with star Alcor) who drove his wagon across the sky as a reward for his devoutness.

POP CULTURE Alcor is the name of a cryogenic freezing company in Arizona. The Big Dipper itself, Mizar and Alcor's home constellation, has made many literary and artistic cameos over the last few centuries. For example, it has appeared in the art and writing of Shakespeare, Homer, Rainer Maria Rilke, and Vincent van Gogh.

13

FOMALHAUT

VISIBILITY ● ● ● ● ● *Fomelhaut A*
              ● ● ● ● ● *Fomelhaut B*
              ● ● ● ● ● *Fomelhaut C*

SEASON Late fall and winter; not visible in Alaska.

QUALITY Very bright, and bluish-white.

DISTANCE FROM EARTH 25 light-years.

DIRECTIONS Fomalhaut lies in Piscis Austrinus and is the "mouth" of the fish that Piscis Austrinus represents. It is the brightest star in its region, and its constellation is adjacent to Aquarius and Capricorn.

SIGNIFICANCE Fomalhaut is one of the brightest stars in the night sky. It is sometimes called "the lonely star," since, in contrast to stars in its vicinity, it shines like a lighthouse over a dim sea. Fomalhaut is another celebrity star, with significance to many cultures: To the Persians, the rise of Fomalhaut meant the beginning of the fall season. When Fomalhaut first appears in autumn, it does not appear very bright—an effect of being near the horizon—though later in winter, as it rises in the sky, its breathtaking brightness becomes apparent. Like most stars low on the horizon, it is sometimes mistakenly identified as red. Do not be fooled—this is merely an effect of Earth's atmosphere, which adds a red tinge to anything low in the sky (even clouds). Higher in the sky, it appears bluish-white.

Fomalhaut inhabits a strange and ruinous solar system. Though quite large and bright, at twice the mass of our Sun, it presides over a debris-ridden region wracked with dust and clouds, not unlike our solar system in its youth. And, indeed, Fomalhaut is a relatively new sun—about one-tenth the age of our own, though with a shorter predicted lifespan.

Far off in Fomalhaut's debris disk sits a lone exoplanet, Fomalhaut B, which gained notoriety for being the first directly imaged exoplanet in history. Fomalhaut B is thought to be about twice the size of Jupiter, but it orbits its home star from a significant distance, farther away even than Pluto is from the Sun. Don't expect to see Fomalhaut B with your telescope, though; it took the Hubble Space Telescope's massive lens to find it in the first place.

When you get Fomalhaut in your scope, it may appear slightly hazy. This is the effect of the debris disk. The next challenge is finding Fomalhaut's siblings. That's right—Fomalhaut is a trinary star, although its companions are quite distant, as they must be in order for planets to form and orbit comfortably. In fact, Fomalhaut B and Fomalhaut C are so far from their parent star, you'll have to completely reorient your telescope to find them.

Fomalhaut B is within the constellation Piscis Austrinus but closer to the next star down in the fish, Delta Piscis Austrinus. About two-thirds of the way down from Fomalhaut A to Delta sits Fomalhaut B, adjacent to the brighter HD 214187. It's a full light-year away from Fomalhaut A and orbits very slowly.

Fomalhaut C, which was only recently discovered to be part of the trinary, is not even in Piscis Austrinus. It is 2.5 light-years away from Fomalhaut A and, in terms of the sky, two degrees north of neighboring star Epsilon Piscis Austrinus, on the other side of the constellation. It is very faint, however—you'll need a map and a bit of luck to pick it out among the background stars.

ORIGIN STORY Though it sounds German, Fomalhaut got its name from Arabic via Ptolemy (who made an early catalogue of constellations). Roughly translated, it means "fish's mouth," and the name of its parent constellation, Piscis Austrinus, means "southern fish."

POP CULTURE Multiple books and stories by science fiction author Ursula K. Le Guin take place on the planets of Fomalhaut. Philip K. Dick's *The Unteleported Man* depicts sending humans to a colony in the Fomalhaut system. Some of Frank Herbert's *Children of Dune* takes place in the Fomalhaut worlds.

RING NEBULA

14

BETELGEUSE

VISIBILITY ● ● ● ● ●

SEASON Most visible on winter evenings after sunset; also visible in the fall and spring.

QUALITY Bright and orangish-red.

DISTANCE FROM EARTH About 500 light-years.

DIRECTIONS Betelgeuse, pronounced *bee-till-joos*, is in the constellation Orion, where it forms Orion's right shoulder (the one holding the arrow).

SIGNIFICANCE Betelgeuse is one of the largest and nearest stars to Earth and belongs to a class of stars called red supergiants. If Betelgeuse were in our solar system, its radius would extend beyond Mars. Massive, diffuse, and red, Betelgeuse is approaching the end of its life cycle, having exhausted its main source of fuel and expanded well beyond its original size. It portends what will happen to our Sun, eventually—although the path that our Sun cuts through the solar system when it becomes a red giant will not be quite as wide as the much-more-massive Betelgeuse.

Betelgeuse is one of few stars for which a **supernova** alert can be issued—that is, there is a chance in our lifetime that it will explode into a supernova, briefly becoming the brightest object in the night sky. Supernovae are rare, especially nearby ones, and when they explode they are briefly the most energetic objects in the universe—even supernovae from distant galaxies are detectable from Earth. Don't hold your breath, though. Scientists estimate that this could happen anytime within the next million years. Telescopic images of Betelgeuse reveal plumes of gas irregularly spewing from the star. It is doubtful that any of Betelgeuse's planets remain, as this giant star has likely destroyed or swallowed them all by now.

Like Fomalhaut's fainter companion Fomalhaut B, Betelgeuse is also a variable star—in fact, one of the most variable in terms of change in brightness. Though the cycle is not regular, it seems to alternate in brightness with a cycle that ranges from one to five years between dimmest and brightest. Careful amateur astronomers, armed with digital tools equipped to measure brightness, can collect data on the precise cycle of dimming and brightening.

There are many sites dedicated to observing and tracking Betelgeuse's varying brightness with a home telescope, some with frequently updated charts showing recent changes. See the Resources section (page 165) for more details on variable star–observing websites and groups. With large telescopes, Betelgeuse appears with "spikes" around it, an effect of lens distortion caused by very bright objects.

ORIGIN STORY The star's name is a French word derived from an Arabic phrase meaning "house of the twins."

POP CULTURE Betelgeuse was first made famous by the Tim Burton–directed comedy *Beetlejuice* (1988), starring Alec Baldwin, Geena Davis, and Michael Keaton (the movie's title is a phonetic pronunciation of the star). Ford Prefect, a character in Douglas Adams's science fiction novel *The Hitchhiker's Guide to the Galaxy*, was from a planet near Betelgeuse.

15

RIGEL

VISIBILITY ● ● ● ● ●

SEASON Most visible on winter evenings after sunset; also visible in the fall and spring.

QUALITY White to bluish-white and very bright.

DISTANCE FROM EARTH About 800 light-years.

DIRECTIONS Below Orion's belt, Rigel forms the lower-left corner of the constellation Orion.

SIGNIFICANCE We continue our tour through Orion's stars by turning to Rigel, the seventh-brightest star in the sky overall. Rigel is a cousin to Betelgeuse in many ways. While also a supergiant and a variable star in Orion, it glows blue rather than red and forms part of a triple-star system. It is also very large in volume—over 50 times the size of our Sun, with 20 times the mass. Eventually, Rigel will turn redder and redder, like its constellation companion Betelgeuse, and then become a supernova, though precisely when is not well determined. When it does, it will, for a few weeks, be about half as bright as a full moon—bright enough to create shadows at night from its light.

Rigel has a binary companion, Rigel B, which can be seen with telescopes of approximately 5 inches or greater. Rigel B is over 300 billion kilometers (186 billion miles) from Rigel A, far enough to remain mostly unperturbed by the violence and mass ejections that accompany a star transitioning to the supergiant phase. Any close planets to Rigel A were almost certainly enveloped when it expanded into a supergiant. Currently, there is no evidence that it hosts any planets of its own, though planet-detecting astronomy is still in its infancy.

Any decent telescope should reveal Rigel B to you upon closer inspection. It appears as a smaller point of light very close to Rigel A. The larger the telescope, the more dramatic the contrast between the pair will appear.

Rigel B is actually a binary as well (Rigel C is the third star's name), though the separation between Rigel C and Rigel B is not visible through a telescopic lens. Both Rigel B and Rigel C are more massive than our Sun and more luminous, too, though they pale in comparison to Rigel A.

ORIGIN STORY *Rigel* is derived from Arabic, originating in a phrase meaning "left foot of the giant." The giant, in this case, is Orion.

POP CULTURE *Invaders from Rigel*, a 1960 science fiction book by Fletcher Pratt, honors this star. Alfred Bester's science fiction novel *The Stars, My Destination* (1956) took place partially in the Rigel system. In 1989, science fiction writer Ursula K. Le Guin and composer David Bedford collaborated on a "space opera" titled *Rigel 9*.

16

ALBIREO

VISIBILITY ● ● ● ● ◉

SEASON Summer and fall.

QUALITY White, but with a small degree of magnification appears as an orange and blue pair.

DISTANCE FROM EARTH About 400 light-years.

DIRECTIONS Albireo is in the Northern Cross. First find Cygnus: The star on Cygnus the swan's face is Albireo.

SIGNIFICANCE Albireo offers a treat to amateur astronomers. The two stars that make up this binary appear as one to the naked eye, yet, with even a small telescope, they separate into two with radically different colors. The yellow, brighter companion, Albireo A, is technically a binary in itself, though it takes a telescope with an enormous aperture to resolve it. Albireo B is a bluish star, slightly smaller than Albireo A. It is not known if the stars A and B are actually gravitationally locked, though it is suspected that they are. If so, they orbit a common point of mass about once every 100,000 years or more. This means their positions relative to each other will not change much in our lifetime.

Albireo B, despite its smaller size, is actually the hottest of the pair—near 12,000 degrees Celsius (21,000 degrees Fahrenheit), more than twice as hot as the main star in Albireo A.

Just to the north and west of Albireo, and adjacent to the next star in the constellation, sits a nebula known as Sh2-91. With an 8-inch telescope (or larger) on a dark night, you should be able to make out this nebula as a diffuse streak of light in the shape of an arc. These are the leftovers of a star that went supernova sometime many thousands of years ago.

The color-differentiated double is actually sometimes easier to see with a weaker telescope. If you are having trouble resolving the two distinct colors, position the star in your viewfinder and then unfocus the telescope slightly. With a little bit of blur, the colors will become more obvious.

ORIGIN STORY The name of this star stems from a mistake. It was first known by the name Ornis, Greek for "bird," and was translated into Arabic and then back to Latin, mistakenly as *ab ireo*, which means "herb." This erroneous translation was mistaken for Arabic and rewritten as *Albireo*, which does not mean anything in Arabic. Today, the Arabic name for this star translates to "hen's beak," which relates to the original Greek meaning.

POP CULTURE A popular flute quartet is named after this star. The 1920 poetry volume by Danford Barney is titled *Chords from Albireo*.

17

POLLUX & CASTOR

SEASON Winter through summer, but best in January through March.

QUALITY Orange (Pollux) and bluish-white (Castor).

DISTANCE FROM EARTH 34 light-years (Pollux); 50 light-years (Castor).

DIRECTIONS The two heads of the twins in Gemini are Pollux and Castor.

SIGNIFICANCE Pollux and Castor are the two twins of the constellation Gemini—which, incidentally, means "twin" in Latin. Though they are very different star systems, they are often mentioned in the same breath. Like any good pair of twins, they are close together in the sky. They are very noticeable as they are the two brightest points in their part of the heavens.

OUR UNUSUAL SUN

You may have noticed a pattern by now. Most of the stars observed in the night sky so far are binaries or greater. Our Sun is actually unusual in this regard, as most stars in the universe are binaries, trinaries, or even larger gravitationally bound systems. It is unknown if life can exist in a binary, as temperatures on orbiting planets would be much more variable with two suns. Multiple-star systems like Castor are the norm—ours is the exception.

Pollux—the more orange of the two—is a giant star that has burned off its hydrogen core and expanded. It has only twice the mass of the Sun, yet nearly 10 times the diameter. Pollux is the brighter of the twins and, in fact, is one of the 20 brightest stars in the night sky.
Part of the reason for this is because it is so close by. Since Pollux is just 34 light-years away from Earth, stray radio broadcasts from our planet have been reaching Pollux for about 75 years. (Yet, as a gas giant, it is unlikely to harbor any alien life forms listening to those radio broadcasts.) There is at least one exoplanet orbiting Pollux, Pollux b, with a mass about twice that of Jupiter. It orbits at nearly the same distance that Mars orbits our Sun.

Castor is a busy sextuple star system. With a decent telescope, Castor appears as a binary, yet each of these stars is actually a binary itself. Additionally, there is a much dimmer binary system orbiting nearby. All three binary pairs have a common center of mass that they orbit.

For our purposes, only Castor A and Castor B are worth referencing, as they are the only two objects visible from your telescope. Castor A's main star, a bright bluish-white star about 50 light-years away from us, possesses twice the mass of the Sun.

Once you locate Gemini, observe both of these stars early in the evening in the winter through early spring. The color differentiation between the two is stunning to behold, and the revelation of Castor as a binary is an added treat.

ORIGIN STORY *Gemini* is Latin for "twins." In Greek mythology, Pollux and Castor were brothers with different fathers—Pollux of the god Zeus; Castor of a mortal. When Castor died in battle, Pollux wept for his brother and asked his father Zeus to let him die, not being able to live without Castor. Zeus granted Pollux the freedom to travel between Olympus, the godly realm, and the underworld, where his twin's spirit lived. It is said that the constellation honors the fraternity between the two.

POP CULTURE Pollux and Castor were the names of two brothers who appeared in *Mockingjay*, a novel in the Hunger Games series by Suzanne Collins. The 1997 John Woo–directed film *Face/Off* featured two terrorist brothers named Castor and Pollux.

WHIRLPOOL GALAXY

18

VEGA

VISIBILITY ● ● ● ● ●

SEASON Best in summer but visible year-round.

QUALITY Very, very bright and bluish-white.

DISTANCE FROM EARTH 25 light-years.

DIRECTIONS Vega is the brightest star in the constellation Lyra.

SIGNIFICANCE Because of its brightness and proximity to Earth, Vega is one of the most studied stars in the sky. Vega was photographed in the 1850s using some of the first photographic equipment, and studying its light emission helped lead to the discovery that stars were made of hydrogen. Tens of thousands of years ago, Vega was the pole star for the Northern Hemisphere, but as Earth's angle of rotation changed, Polaris (see page 47) became the pole star known today. In about 12 millennia, as Earth's angle of rotation shifts further, Vega will regain its position as our northern star.

Vega has about twice the mass of our Sun, yet it is much, much brighter—in fact, it is the fifth-brightest star in the sky overall and the second-brightest in the Northern Hemisphere.

Vega forms part of what is called the "Summer Triangle"—a juxtaposition of three stars from three different constellations, Vega, Deneb, and Altair, which together form a triangle. Deneb is in the constellation Cygnus, and Altair is in Aquila. The three frame a portion of the long band of the Milky Way, visible on dark nights, though if you live in a dark enough region to see the Milky Way, you can see Vega regardless of whether you can find the Summer Triangle.

In the summer months, the Summer Triangle trio rises with sunset and sets around dawn. In this time frame, Vega appears through a telescope as very brilliant and bright. A cloud of dust and debris surround the star, possibly the result of a planetary collision. The debris disk produces a slight hazy ring around the star when viewed with excellent resolution.

Vega's home constellation, Lyra (which means "harp"), is host to some other interesting stars. On the opposite side of Lyra sit Sheliak and Sulafat, two stars that form points of the harp. Between these two, about two-thirds of the way toward Sheliak, is the Ring Nebula—a very famous nebula that resembles, fittingly, a ring. With a 3-inch reflector the Ring Nebula appears as a faint hazy circle; larger telescopes can see its greenish color. With a very large telescope you may be able to see the central star from which the nebula's gas and dust likely emanated in a massive explosion thousands of years ago.

ORIGIN STORY The name *Vega* is a transliteration from Arabic that means "falling eagle."

POP CULTURE Isaac Asimov's Foundation series of science fiction novels features the Vega system as a prominent province. In the novel *Contact* by Carl Sagan, and in its movie adaptation, alien radio transmissions are traced to Vega. The television show *Babylon 5* featured the "Vega Colony" as a prominent settlement.

19

ALDEBARAN

VISIBILITY ● ● ● ● ●

SEASON Fall, winter, and early spring.

QUALITY A very bright, red star.

DISTANCE FROM EARTH 66 light-years.

DIRECTIONS Aldebaran is the brightest object (the eye) in the constellation Taurus. Orion's belt also points, roughly, to Aldebaran.

SIGNIFICANCE One of the most visible stars in the night sky, Aldebaran is often used as a starting point for finding other, more difficult-to-find objects. Once you can find Aldebaran and some of the other bright ones like Sirius, finding fainter nebulae and galaxies becomes easy.

Like many of the brightest stars in the sky, Aldebaran is an orange giant. This means it has reached the end of its long life and has expanded well beyond its original diameter as it runs out of hydrogen fuel, becoming vastly larger, brighter, and redder. Through telescopes, it has a very bright and jewel-like appearance, with a tint more orange than our own Sun. Some measurements seem to suggest that Aldebaran has at least one orbiting planet, but this has not been confirmed.

Aldebaran is one of the few bright objects that is occulted, or hidden from view, by the Moon. Similar to an eclipse, **occultation** occurs when the Moon passes in front of a star. This creates a brief but exciting moment when the star is only partially occulted, which allows astronomers to simultaneously observe specific features on the Moon that are lit by the star as well as specific features of the star—for instance, how wide it is (which can be calculated knowing how long it takes to occult). Aldebaran occultations occur over 3-year time spans separated by approximately 19 years. The current cycle has about 49 occultations (like solar eclipses, visibility varies depending on your location on Earth). Aldebaran's current occultation cycle lasts from 2015 through 2018, and the next series begins in August 2033 and ends in February 2037.

ORIGIN STORY In Arabic, *al debaran* means "follower," and this planet it so named because it follows the Pleiades (see page 111) in the sky.

POP CULTURE Aldebaran is a popular star in fiction. Authors who have mentioned the star in their works include Ursula K. Le Guin, Alfred Bester, J. R. R. Tolkien, George Orwell, James Joyce, H. P. Lovecraft, Thomas Hardy, and Douglas Adams.

20

ARCTURUS

VISIBILITY ● ● ● ● ●

SEASON Spring and summer.

QUALITY Bright and slightly reddish.

DISTANCE FROM EARTH 37 light-years.

DIRECTIONS In the constellation Boötes, Arcturus forms the lower vertex or the left knee, depending on the drawing.

SIGNIFICANCE Arcturus is one of the brightest stars in the Northern Hemisphere; in most of the Northern Hemisphere, only Sirius (see page 53) shines brighter. As an orange giant, Arcturus is toward the end of its life and will contract to a white dwarf eventually as the outer gaseous layer of the star blows off into space. It is about the same mass as our Sun, only a few billion years older, and foreshadows the fate of our solar system.

Arcturus is another celebrity star, particularly because of its historical significance. Many cultures associated its rise with seasonal change. The Romans thought the star was a predictor of bad weather.

At the start of the 1933 World's Fair in Chicago, the lights to inaugurate the fair were activated by focusing the light of Arcturus, via a telescope lens, onto a photocell (a basic solar panel), which tripped a current and turned on the switch.

If you were to observe it carefully from year to year, you might notice that Arcturus is one of the fastest-moving stars in view. This property is called **proper motion**—the speed at which stars appear to move around the sky, relative to our view on Earth. In 1,800 years, Arcturus will move 1 degree across the sky—which is fast for a star. (The star with the fastest proper motion is Barnard's Star, a red dwarf 6 light-years away that moves 1 degree across the sky in 350 years).

Look for Arcturus in the evenings during summer, around 8 p.m. to 10 p.m. The last two stars in the handle of the Big Dipper make an "arc to Arcturus"—a common mnemonic for remembering how to find Arcturus. (The full mnemonic is actually "make an arc to Arcturus and speed on to Spica." If you continue to follow this imagined arc, you will end up looking at Spica in the Virgo constellation.)

Arcturus will appear reddish in contrast to the other stars in its vicinity. It is not currently known to have any planets or smaller orbiting stars. Enjoy Arcturus while it lasts. It is moving away from us so fast that in about a million years it will no longer be visible from Earth.

ORIGIN STORY This star's name is Greek in origin and means "guardian of the bear"—a reference to the close proximity of its home constellation Boötes to Ursa Major and Ursa Minor ("Greater Bear" and "Lesser Bear").

POP CULTURE Arcturus appears in certain English translations of the Bible, in Job 9:9: "Which makes Arcturus, Orion, and Pleiades, and the chambers of the south." In Roman playwright Plautus's *Rudens*, from the third century BC, the personified star recounted the prologue.

21

CAPELLA

VISIBILITY ● ● ● ● ●

SEASON Fall and winter.

QUALITY Bright and yellowish-white.

DISTANCE FROM EARTH 42 light-years.

DIRECTIONS In the Auriga constellation, Capella is the brightest star.

SIGNIFICANCE Just beyond Gemini and Taurus sits unassuming Auriga, an ancient constellation named by Ptolemy. Auriga is a planet-rich constellation—six stars within its bounds are known to house a planetary system, though most of them are faint, hard-to-see stars with dry names like HD 43691 and HD 45350. (As you may have surmised, no exoplanets are visible with any kind of home telescope—in fact, only a few exoplanets can be directly imaged even with large scientific telescopes.)

Capella appears as one star but is actually four. You heard that right: Capella's brightness represents the light emanations of four different stars, which are actually two tight binaries that orbit around a common point. These two binaries, Capella A and Capella B, couldn't be more different. The two stars of Capella A have about the same mass, at around 2.5 times the mass of our Sun. They're both much, much brighter than the Sun and orbit each other in about 100 days, meaning they're closer to each other than the Earth is to the Sun. Capella B is a binary of two faint red dwarfs that orbit Capella A from trillions of miles away, making a couple of revolutions per millennium.

When looking through your telescope, don't try to see Capella as a binary, though—you can't. The stars of Capella A are simply too close, and the stars of Capella B are simply too dim, to resolve as separate points of light. However, Capella is, by virtue of brightness, one of the few stars you can observe in the daytime. You'll have to measure its position very precisely, as guide stars and constellations will not be available to you. And as always, make sure you are facing nowhere near the Sun, as even a glance through an unfiltered telescope at the Sun can blind you. Viewing Capella under a blue sky is an entertaining challenge, to say the least.

Just off to the side of Capella sits Haedi, also known as "the Kids"— a triangle of three stars that form a small cluster. If you're wondering whether you are indeed looking at Capella, just look for this tiny adjacent cluster.

ORIGIN STORY *Auriga* is Latin for "chariot rider," while *Capella* means "small goat." *The Kids* refers to the diminutive name for a baby goat—a kid. Capella was supposed to represent the goat from which Zeus nursed as an infant. Though it looks very similar, the Italian term *a cappella*, which references unaccompanied singing, has nothing to do with this star system.

POP CULTURE Philip José Farmer set his book *Time's Last Gift* in this star system. Anne McCaffrey's series of novels titled The Tower and the Hive takes place partly in the Capella planetary system.

22

ALGOL

VISIBILITY ● ● ● ● ◐

SEASON Autumn, around Halloween.

QUALITY Bright, white, with changes in brightness (becoming dimmer) every few days.

DISTANCE FROM EARTH 93 light-years.

DIRECTIONS Find Perseus in the sky—Cassiopeia points at it. Algol is the bright star along the line of five stars that extend out from Mirfak in a hook shape.

SIGNIFICANCE Can a star be possessed? For centuries, Algol has been called the Demon Star. This is not because its scorching surface burns condemned souls, but rather because the star exhibits an affinity for spontaneously becoming much dimmer.

Happily, Algol's sudden dimming and brightening is not attributable to witchcraft, but rather to the fact that its point of light conceals three stars, one of which (Beta Persei B) periodically passes in front of the brightest (Beta Persei A). When Beta Persei B eclipses its sibling, the appearance from Earth is of a sudden drop in brightness, as the brighter star is concealed. Indeed, Beta Persei A is around 30 times brighter than Beta Persei B. This effect of a star concealing some of a neighboring star's light is known as occultation (as seen with Vega on page 72), which also plays in to Algol's demonic theme. Likewise, this star is rumored to be most visible (eerily) on Halloween.

The occultation of Beta Persei A is so regular that the duration is known down to the second. Every 2 days, 20 hours, 48 minutes, and 57 seconds, Algol will start to dim, soon becoming three times dimmer than it looked a few hours prior. This occultation will continue for 9 hours and 40 minutes, and then it will return to its normal brightness. You can check online for the exact day and time that the occultation starts and plan to observe during a time when it starts in the evening. If you've got Algol lined up in your telescope right as it starts, you should see it very clearly "wink" and dim. The dimming takes a total of five hours, and then for the next five hours it slowly brightens again as the occultation ends.

The Demon Star is supposed to form the eye of Medusa in certain constellation drawings. You may have seen the Perseus constellation depicted as a sword-carrying man with winged feet (that's Perseus) waving his blood-soaked sword with his right hand and holding the severed head of Medusa in his left. As with most constellations, it can take a lot of imagination to connect Perseus's stars and see the figure depicted by ancient civilizations.

ORIGIN STORY *Algol* means "demon's head" in Arabic.

POP CULTURE ALGOL, an acronym for algorithmic language, was one of the first computer programming languages. The 1920 German film *Algol: Tragedy of Power* was about extraterrestrials from this star system.

MIRA

VISIBILITY ● ● ● ● ● Varies.

SEASON Highest in summer; also visible in fall and winter.

QUALITY Orangish-white, with peak brightness as noted below.

DISTANCE FROM EARTH 420 light-years.

DIRECTIONS Mira is in the constellation Cetus, the whale. A long line of stars connects the two sides of the whale, its body, and fins. Mira is between these two, specifically between Delta Ceti and Zeta Ceti.

SIGNIFICANCE Algol may be the Demon Star, but Mira is arguably stranger. Not only does this star flicker on and off, and in a more dramatic fashion than Algol, but it also has a 13-light-year-long tail! While it is unknown if the tail has a demonlike fork, it has certainly been growing for a long time. As Mira speeds through space, it leaves a wake behind it not unlike a boat through the water or a jet through the sky. Unfortunately, the comet-like tail is not visible except in the ultraviolet band, and ultraviolet light is filtered out by Earth's atmosphere. You'd have to be above the atmosphere to see it clearly.

Mira's behavior as a variable star is curious. It is visible with the naked eye for about four months, and then it becomes too dim to see unaided for seven months, before slowly starting to brighten again. The periods of maximum brightness occur every 11 months, with some regularity—yet observers have noticed its period shift slightly over the years. The year 2011 was the brightest year on record for Mira. In 2012 it was significantly dimmer at its peak brightness. The following dates are the expected maximum brightness periods for Mira (though it should be noted that Mira is sometimes fickle and these may vary):

MIRA'S MAXIMUM BRIGHTNESS

April 20–April 30, 2016

March 21–March 31, 2017

February 19–March 1, 2018

January 20–January 30, 2019

December 21–December 31, 2019

November 20–November 30, 2020

October 21–October 31, 2021

September 21–October 1, 2022

August 22–September 1, 2023

July 22–August 1, 2024

June 2–July 2, 2025

May 23–June 2, 2026

April 23–May 3, 2027

March 23–April 2, 2028

February 21–March 3, 2029

January 22–February 1, 2030

December 23, 2030–January 2, 2031

November 23, 2031–December 3; 2031

Beyond 15 years, it becomes difficult to predict Mira's tendencies with accuracy of more than a week. Note that the days listed are the periods of maximum brightness. Mira will start to brighten about two months prior and continue to dim for another couple of months before becoming invisible to the naked eye.

What is the origin of Mira's strange behavior? The means by which Mira brightens and darkens is not wholly understood. It is suspected, however, that Mira has a white dwarf companion that periodically sucks the mass from Mira's outer layers. This occurs in a regular pattern and causes the predictable dimming and brightening.

The word *Mira* has become the general term for a class of variable stars with similar properties. The American Association of Variable Star Observers (aavso.org) is dedicated to observing and tracking the Mira-class variables (as well as other variable stars).

ORIGIN STORY *Mira* is Latin for "astonishing" or "wonderful," reflecting the mood of the Renaissance astronomers who observed and recorded its variable tendency. (Pity poor Algol, derided as a "demon" for its variability, while similarly variable Mira is heaped with praise.)

POP CULTURE Mira has been featured in several episodes of the original *Star Trek* television series and *Star Trek: The Next Generation.* A romance publisher is named Mira Books.

OMEGA NEBULA

24

PERSEUS DOUBLE CLUSTER

VISIBILITY ● ● ● ● ●

SEASON Spring.

QUALITY Two adjacent star clusters appearing as dense fields of stars through a telescope.

DISTANCE FROM EARTH 6,800 to 7,500 light-years.

DIRECTIONS Find Perseus and its neighbor Cassiopeia. Cassiopeia is shaped like a *W*—the lower vertex of the first *V* of the *W* is Ruchbah. The nearest point of light to Ruchbah in the constellation Perseus is Eta Persei. The Double Cluster appears about one-third of the way from Eta Persei toward Ruchbah. It looks like a single star to the unaided eye. Alternatively, look for Perseus's sword—the Double Cluster is on the sword's handle.

SIGNIFICANCE The Double Cluster is one of the gems of the sky, and it is easy to see why. It is the first object in our guide that is not just a few orbiting stars, but a whole field of stars. A **cluster** is a very large group of stars that are gravitationally bound—they orbit in a swarm for eternity, unless they are forced apart by a close encounter that flings one out of orbit. Often, stars within clusters are formed from the same interstellar clouds of gas: In other words, they are of a common kin. Clusters contain hundreds or even thousands of stars that float together and weave around each other in a dazzling cosmic dance.

The Double Cluster appears in the sky as two separate clusters, but the two objects, known formally as NGC 884 and NGC 869, could be gravitationally bound. Their combined mass is over 10,000 times that of our Sun. They are the same age, around five million years old, which is quite young in galactic terms (in comparison, our planet and Sun are 4.5 billion years old). They are currently moving closer toward Earth and will continue on that path for some time before retreating.

The longer you stare at the Double Cluster, the more stars you're apt to see—so many, indeed, that it is difficult to focus on a single point of light. On a bright night, say, with a fully lit Moon, the light pollution might limit your vision to a few dozen stars. But on a dark night, the number will be truly staggering. If you have a camera on your telescope, or are content to use your smartphone's camera to snap a picture through the eyepiece (see page 21), download the photograph and count the stars on your computer screen. Try this under different conditions (higher and lower elevations, more or less light pollution) for comparison.

ORIGIN STORY For millennia, this object was known as one star. It wasn't until the early nineteenth century that astronomer William Herschel recognized it as two clusters and coined its name.

POP CULTURE James A. Michener, in his novel *Space*, described the Double Cluster as "a staggering collection of great stars engaged in some kind of combat." A "double cluster" is also a kind of crochet stitch.

25

HERCULES GLOBULAR CLUSTER

M13

VISIBILITY ● ● ● ● ●

SEASON Visible all year; best in summer and fall.

QUALITY A fuzzy dot with the naked eye; a rich cluster of stars through a telescope.

DISTANCE FROM EARTH 25,000 light-years.

DIRECTIONS First, find Hercules, a constellation that appears between Arcturus (see page 76) and Vega (see page 72). The center of Hercules is bound by a trapezoid. The Hercules Globular Cluster is between Eta and Zeta Herculis, or on the Arcturus side of the trapezoid. If you look halfway between these two stars in the trapezoid, you will see a faint fuzzy region. Zoom in on these with your telescope to see the Hercules Cluster in its full brilliance.

SIGNIFICANCE The Hercules Globular Cluster is one of a few systems to which humans have sent radio signals in hopes of contacting other intelligent life. The Arecibo Observatory in Puerto Rico sent a message toward this cluster in 1974. Written by astronomers Frank Drake and Carl Sagan, the message encoded information about humans, mathematics, and biology. Since the Hercules Globular Cluster is 25,000 light-years away, it will take that long for the message to arrive and, if anyone is listening, at least the same length of time for a response to come back.

The Hercules Globular Cluster, also known as M13, was chosen as a candidate for communication attempts partially because of the sheer enormity of stars in the region—estimated to be in the range of hundreds of thousands. That is indeed a large number but pales in comparison to the number of stars in the galaxy, which is around 300 billion. It is possible, of course, that there could be an intelligent alien somewhere among those hundreds of thousands of stars.

The Hercules Globular Cluster is visible without a telescope, but just barely. If you are looking through binoculars, you might not be able to see it clearly as a cluster—in fact, many binocular stargazers have mistaken it for a comet. Through your telescope, the Great Cluster, as it is also known, reveals its full self, resembling a pointillist splat of paint, like stars of light radiating from a target. If you are using a telescope with a very wide reflector, you will never be able to count all the stars—there are simply too many, thousands and thousands visible even from your backyard. Hubble images of M13 are particularly impressive, as they capture the distinct color differences between blue and yellow stars in the cluster.

ORIGIN STORY The cluster is named after its home constellation Hercules, which nods to the mythological figure who was Zeus's son and battled with giants.

POP CULTURE The most famous quote about the Hercules Globular Cluster, also known as Messier (pronounced *messy-a*) object 13 (or M13), comes from American author Kurt Vonnegut, who wrote on the dedication page of *The Sirens of Titan*: "Every passing hour brings the Solar System forty-three thousand miles closer to Globular Cluster M13 in Hercules—and still there are some misfits who insist that there is no such thing as progress."

26

DUMBBELL NEBULA

VISIBILITY ● ● ● ● ●

SEASON Summer.

QUALITY A circular, bright blur; blue with red edges and much larger than nearby stars.

DISTANCE FROM EARTH 1,400 light-years.

DIRECTIONS While the Dumbbell Nebula is dim (a **nebula** is a cloud of gas and dust, often a remnant from an exploded star), it is easy to find if you can identify other stars in the region. The brightest nearby star is Rotanev, in the Delphinus constellation. Next, locate Albireo (see page 66). Along this line between the two stars, two-thirds of the way toward Albireo, you'll find the Dumbbell Nebula. Alternatively, if you can find the faint constellation Vulpecula (adjacent to Albireo) and neighboring faint constellation Sagitta, you will be able to see Dumbbell Nebula sitting between the two, near the bottom of the middle star in the M shape formed by Vulpecula.

SIGNIFICANCE The Dumbbell Nebula is the aftermath of a dying star that expanded and shed its core some 15,000 years ago. Shaped vaguely like a dumbbell as viewed from Earth, the luminous remnants resemble the future of our own solar system. Once the Sun ceases to fuse hydrogen in its core, in four billion years or so, it will expand, blow off its outer layers, and explode into a vast sphere of ever-expanding dust and gas.

The Dumbbell Nebula is famous for being one of the first nebulae ever discovered, by Charles Messier in the eighteenth century. In the late Renaissance, advances in optics technology led to the discovery that many objects in the sky were not mere points of light, but hazy cloudlike objects. The word *nebula* actually means "cloud-like" or "vaporous."

The shape of the Dumbbell Nebula is apparent with even a small telescope. With telescopes greater than 6 inches, the colors and luminance are more visible. You may even be able to see a tiny star at the center of the nebula: This star, which has about half the mass of our Sun, is all that remains of this once-giant sun. It is about 15 times hotter than our Sun, though, at 85,000 degrees Celsius (153,000 degrees Fahrenheit).

The Dumbbell Nebula displays colors that are visible with a good telescope. The outer part of the familiar hourglass shape appears as red, which is the resulting color of heated hydrogen. The oval halo appears as more of a bluish color, the color of heated oxygen, sulphur, and nitrogen—all of which were ejected when the original star exploded.

ORIGIN STORY This nebula is so named because of its resemblance to a dumbbell, although it is often cited as looking more like an hourglass. A smaller nebula, the Little Dumbbell Nebula, is named for its similarity in appearance.

POP CULTURE Poet Steve Kowit has a poetry collection named for the Dumbbell Nebula.

27

BEEHIVE CLUSTER

VISIBILITY ● ● ● ● ●

SEASON Late winter to spring.

QUALITY Appears as a small "swarm" of stars.

DISTANCE FROM EARTH 500 light-years.

DIRECTIONS Find the constellation Cancer, one of the zodiacal constellations. Cancer is shaped more or less like a *Y*. You will find the Beehive Cluster at the bottom of the *Y*, between the vertex (Asellus Australis) and the next star down (Asellus Borealis) and a little off to the side.

SIGNIFICANCE The Beehive Cluster has fascinated observers and astronomers for thousands of years. The ancient Greeks recorded it in their poetry, and in the second century BC, Greek astronomer Hipparchus of Nicaea included the cluster in his star catalogue. After the telescope was invented it was one of the first studied objects, and Galileo famously observed it.

The Beehive Cluster is an open cluster, meaning all stars within it were formed from the same cloud of dust and gas. It is estimated that it formed around half a billion years ago, and that about 1,000 stars are bound in the cluster. Three hundred of these stars are similar in composition and size to our Sun. However, the most visible stars from our vantage point on Earth are the sparse few red giants, which have expanded to thousands of times the size of our Sun and glow orange. In good viewing conditions these will appear yellowish in contrast to surrounding stars.

Unusually, the Beehive Cluster looks more interesting with binoculars or a smaller telescope than with a big telescope. If you view it with a big telescope, you risk seeing a small, uninteresting patch of stars rather than the magnificence of a large, dense cluster.

Several of the stars in the cluster are known to have their own planets. These detected planets are of a class known as hot Jupiters—Jupiter-like worlds that orbit much closer to their parent stars than our distant Jupiter does to the Sun.

By watching how this star cluster moves, astronomers determined that it seems to have a relationship with the Hyades (see page 95). In fact, the two clusters probably were once the same mega-cluster, and the two different gravitationally bound groups drifted away from each other over the course of millions of years.

ORIGIN STORY Before it was known to be a cluster of stars, the Beehive was known to the ancient Greeks as Praesepe, which means "manger." Two neighboring stars, Asellus Australis and Asellus Borealis, were sometimes depicted as donkeys that ate from the manger. It is now called Beehive because of its resemblance to a swarm of bees.

POP CULTURE NASA scientist and science fiction writer Geoffrey A. Landis's *Embracing the Alien*, a "space opera," takes place in the Beehive Cluster. Aratus, an Ancient Greek poet born in the fourth century BC, wrote of the Beehive Cluster in his poem *Phaenomena*: "Watch, too, the Manger. Like a faint mist in the North it plays the guide beneath Cancer."

28

HYADES CLUSTER

VISIBILITY ● ● ● ● ◐

SEASON Winter and spring.

QUALITY Appears as multiple stars in the night sky, with many more visible though a telescope.

DISTANCE FROM EARTH 150 light-years.

DIRECTIONS The Hyades are in the Taurus constellation. The V-shaped head of the bull Taurus is entirely composed of stars from the cluster. They sit next to Aldebaran (see page 74), such that Aldebaran appears to follow them as they move through the sky.

SIGNIFICANCE The Hyades and the Beehive Cluster (see page 93) have similar origins—though they have broken off into two separate groups. The two likely formed from the same cloud of dust and gas into one large cluster, which then became unbound and separated millions of years ago.

Many of the other clusters in this book appear as faint, fuzzy stars in the sky, which, upon closer inspection, become clusters of thousands of stars. However, the Hyades are so close to Earth that it appears simply as a number of stars in Taurus—save Aldebaran, which is unrelated to the Hyades and just happens to appear in the same field of view.

With the naked eye, the Hyades appear as 5 to 10 stars, depending on light pollution. Yet with a telescope, the Hyades is revealed to contain dozens, even hundreds of stars. In total, the Hyades are composed of about 400 stars. The few you see with the naked eye just happen to be the brightest and biggest, and all the brightest are gas giants or red giants.

The Hyades are notable for being the closest open cluster to Earth (and therefore the most studied). They have existed for more than half a billion years and, because of its situation in an emptier part of the galaxy, may exist as a cluster for hundreds of millions of years to come. The continued existence of this cluster of stars is a bit of a cosmic mystery: Most clusters disperse within 100 million years, and it is thought that only by virtue of its staggering mass—about 1,000 times the mass of our Sun—that the Hyades have remained gravitationally bound after 600 million years.

ORIGIN STORY In ancient Greek mythology, the Hyades were a group of sisters—daughters of Atlas and siblings of Hyas, an archer killed while hunting. Distraught at Hyas's death, the sisters wept for him, and the appearance of the Hyades Cluster is said to represent their tears. The connection to rain is noteworthy.

POP CULTURE The Hyades appeared in Homer's *Iliad*. One notable appearance of the Hyades is in the poem "Ulysses" by Alfred, Lord Tennyson: "Greatly, have suffered greatly, both with those / That loved me, and alone; on shore, and when / Through scudding drifts the rainy Hyades / Vexed the dim sea." The phrase *rainy Hyades* refers to the rise of Hyades in the sky, which occurred during the rainy season. The Hyades are also called the "April Rainers" in English folklore.

29

IRIS NEBULA
NGC 7023

VISIBILITY ● ● ● ● ●

SEASON Year-round.

QUALITY A bluish-white cloud with a central point of light.

DISTANCE FROM EARTH 1,400 light-years.

DIRECTIONS The Iris Nebula is just off the side of Cepheus, between Beta Cephei and Alpha Cephei (though closer to Beta).

SIGNIFICANCE The Iris Nebula is a difficult nebula to sight, but only slightly—binoculars are sufficient to reveal it. It is barely visible with the naked eye, but careful viewing reveals its cloudy, murky starlight reflected through massive clouds of gas. The central star illuminating this nebula is huge, approximately 10 times the mass of our Sun, and has the rather drab name of SAO 19158.

This kind of nebula is known as a reflection nebula. This means that the glowing gas effect is caused by the refraction and scattering of light from a star, which illuminates the gas and dust clouds and heats them

97

such that they glow, too. The visual effect—of a central point of light illuminating a surrounding cloud of gas—is similar to how a fluorescent light works.

The Iris Nebula actually contains a number of objects. The star cluster within the nebula is called NGC 7023 and sits just beyond the glowing region of the nebula. It is particularly visible because the nebula blocks background stars with its stray gas and dust. The lack of background stars should help you be certain you're looking at the right nebula.

If your telescope has the capacity to take a long-exposure photograph, the Iris Nebula appears as a bluish-white smear around a central star. The contrast of the bluish-white smear on the dark background sky is very dramatic and is indicative of the sheer size of the nebula— 6 light-years from end to end.

The Iris Nebula is a more recent discovery and was imaged first by William Herschel in the late eighteenth century. Recent Hubble images have revealed the nebula in stunning depth. Studies of the nebula have revealed strange organic molecules in the vicinity of the nebula, of a common type of carbon-hydrogen particle present in many exhaust fumes. These particles tint the nebula red, though with Earth's atmosphere, the faint reddish color is hard to make out and the nebula generally appears bluer.

As it is near the northern celestial pole, the Iris Nebula should be visible year-round during the nighttime. Finding it first with the naked eye is much easier in areas with low light pollution, though if the general region can be discerned, you should be able to locate it regardless.

Nearby Beta Cephei may also be of interest to amateur astronomers interested in variable stars: It has a very slight variability that goes from maximum to minimum every 4 hours and 34 minutes.

ORIGIN STORY The Iris Nebula's namesake is the purple flower, the iris. To most, though, it looks more like a glowing eye than a flower.

POP CULTURE By virtue of its relative faintness and the recent discovery of its beauty, the Iris Nebula has few pop culture references. It appears in the lore of the 2000s video game *Halo*.

BETELGEUSE AND
THE ORION NEBULA

30

OMEGA NEBULA

VISIBILITY ● ● ● ● ●

SEASON Summer and fall.

QUALITY Not visible to the naked eye; appears as a glowing cloud with a number of stars in its center.

DISTANCE FROM EARTH 6,000 light-years.

DIRECTIONS The Omega Nebula is next to the constellation Sagittarius. Locate Sagittarius in the sky, which resembles either a newspaper hat or a teapot when stars are connected. Imagine a line between the two stars Kaus Australis and Kaus Media, and then follow the line upward for another whole width of Sagittarius. The Omega Nebula is only a few degrees from the Eagle Nebula (M16; see page 119), and both can be viewed together easily on a summer evening.

SIGNIFICANCE The Omega Nebula goes by many different names, including the Horseshoe Nebula, M171, and the Swan Nebula. For simplicity's sake, Omega is used here. It is a huge and massive nebula, about 1,000 times as massive as our Sun and 15 light-years across. One thousand stars shine within the Omega Nebula. It is a relatively new nebula, thought to be about a million years young.

It is also a crèche for young suns: Stars are constantly being created out of Omega's gaseous fields and will continue to be birthed there for a long time yet.

Depending on the magnification of your telescope, you may see the central cluster of stars in this nebula as being U-shaped. Indeed, the name *Omega Nebula* originates in an early description of its shape as resembling the Greek letter omega, Ω. With a small telescope, it will look more like a *U*. If you have a telescope that is 8 inches or larger, the "hooks" at the bottom of the *U* will make it appear more like an omega. Larger lenses reveal knots and details of the gaseous structure, along with more clustered stars.

There are a number of clusters and nebulae in this region of the sky because it is near one of the bands of the Milky Way Galaxy, the central plane where its arms are centered. As noted previously, just off to the side of the Omega Nebula sits the Eagle Nebula, similar in size and composition. When viewed through binoculars, these two might even appear in the same image. Just down from the Omega Nebula and closer to Sagittarius is the Trifid Nebula (see page 105) and the Lagoon Nebula (see page 121), its neighbors in the sky.

ORIGIN STORY This nebula's various names are a reflection of what it resembled when first seen through a telescope. The shape has been variously described as a horseshoe or an omega (Ω), its namesake. It was one of the objects classified by Charles Messier in his 1771 star catalogue; hence one of its designations is Messier 17 or M17.

POP CULTURE The Omega Nebula appears in Poul Anderson's 1970 science fiction novel *Tau Zero*, and it also appears in the 2011 video game *Mass Effect 2*.

31

WILD DUCK CLUSTER
M11

VISIBILITY ● ● ● ● ●

SEASON Evenings in summer and fall.

QUALITY Appears as a tightly packed cluster of stars, denser in the center and sparser as you look farther away. It is said to resemble a flock of birds.

DISTANCE FROM EARTH 5,000 light-years.

DIRECTIONS The Wild Duck Cluster is in the constellation Scutum (the shield). Start from the Aquila constellation, which is readily visible because of the presence of bright star Altair. Aquila resembles a kite with a tail. Just next to the second-to-last star on the kite's tail, find the Wild Duck Cluster.

SIGNIFICANCE Only 250 billion years young and teeming with bright blue stars, the Wild Duck Cluster offers a snapshot of an adolescent portion of the Milky Way. It is one of the densest and most tightly bound clusters visible from Earth. In this tiny section of sky, barely visible with the naked eye, almost 3,000 stars are clustered together in a region that spans 20 light-years. Consider that for a second. In our neighborhood, there are about 0.004 stars per cubic light-year, and the nearest neighbor, Alpha Centauri (see page 51) is about 4.4 light-years away. The Wild Duck Cluster has about 1.5 stars per cubic light-year. In other words, it is around 400 times as dense with stars. If our planet were in the Wild Duck Cluster, the sky would be so bright that one's shadow would easily be seen at night.

The Wild Duck Cluster is slightly harder to find than some of the previous nebulae, though no less rewarding. Fortunately, it is situated close enough to the Aquila constellation to make it readily visible. It can be helpful to start with a pair of binoculars before transitioning to a telescope, as binoculars can easily find the "kite tail" section of Aquila, which will help you locate the cluster quickly. With a telescope it looks far richer and more detailed than with binoculars, however.

Try viewing the Wild Duck Cluster with a smaller scope and slowly increasing magnification. You'll find that the Wild Duck Cluster is somewhat like a fractal in the sense that it becomes richer and denser the closer you look. There is really no limit to its density, no matter how much you zoom in; even the most powerful telescopes will not be able to reveal all its mysteries. There are certainly many unnamed and undiscovered stars hiding within.

In fact, if you have the capability of taking a long-exposure photograph with a camera attachment, you'll see the number of stars increase even further. There are many dim stars in the cluster that are not visible through your telescope but that will be revealed with longer exposures.

ORIGIN STORY This cluster is so named because it resembles a cluster of wild ducks flying across the sky.

POP CULTURE New Zealand artist John Reynolds created an abstract artwork entitled *Wild Duck Cluster,* which is in the collection of the Chartwell Museum in Auckland, New Zealand.

32

TRIFID
NEBULA

VISIBILITY ● ● ● ● ●

SEASON Summer and fall.

QUALITY A few stars surrounded by a red and blue cloud, with dark rivulets running through.

DISTANCE FROM EARTH 4,400 light-years.

DIRECTIONS Find the familiar "teapot" of Sagittarius. On the side of the teapot facing Scorpio—that is, the spout side rather than the handle side—locate the scalene triangle formed by the three stars of the spout. If you imagine a line from Kaus Australis (the star at the pot's base) directly between the other two stars in the triangle, it will touch both the Lagoon Nebula and the Trifid Nebula. It's just about twice as far from the bottom to the top of the triangle as it is to the two nebulae.

SIGNIFICANCE With the best telescopes, the Trifid Nebula appears as a red crystal ball, with a dark vein running through it (a cloud of gas and dust) and blue-green wisps emanating from one side like steam through a field of stars. If it sounds dramatic, it is. The Trifid Nebula is the result of a violent cosmic drama. It is a mix of a star cluster and a nebula of hot gas and dust, with the different visible colors emblematic of different kinds of atoms being lit. This huge nebula is not so different from the one that spawned our Sun, only it is much younger—about 300,000 years old, meaning only a little older than humans (*Homo sapiens*) as a species.

Unfortunately, many of the stars in the Trifid Nebula are born for a short life. Because of the massive gravity of the nebula's innards, many new stars are pulled back into danger shortly after their birth, at which point they again explode or shed their skin as their mass increases proportionally. Slowly, this nebula sheds excess gas into the void of the Milky Way, where it will never again form a star. Eventually, the nebula will stabilize and its newborn stars won't have such a short, fleeting existence.

The Trifid Nebula has many strange features, some of which are visible with a home telescope and some of which can be seen only though scientific telescopes like the Hubble. One famous region of the Trifid looks vaguely like a unicorn horn, though it is really a pillar of gas and dust.

The Trifid Nebula is barely visible with the naked eye, but striking when magnified with a telescope. Once you've located it in the sky, what you see may depend on light conditions and your telescope's strength. At the minimum, you should see two or three bright stars surrounded by two different spheres of color, one blue and one red. The red one is laced with a vine of dark dust running through it. The richness of colors may be increased in darker regions; similarly, with a high-powered telescope, you will certainly see more.

From here, it's just a quick hop over and up to the Lagoon Nebula (see page 121), which, befitting its name, is slightly more placid in comparison to the Trifid.

ORIGIN STORY The word *trifid* is Latin for "threefold." The name references the nebula's appearance as having three distinct nebulae within: a blue one, a red one, and a very dark one, representing different temperatures and gaseous constituencies.

POP CULTURE The magazine *Cracked* called the Trifid's unicorn horn one of the "most bizarre things ever discovered in space." This nebula was referenced in the original *Star Trek*, and it makes an appearance in Poul Anderson's 1970 science fiction novel *Tau Zero*.

M4
MESSIER 4

VISIBILITY ● ● ● ● ◐

SEASON Summer.

QUALITY Looks like a single star in the sky, and with light magnification takes on a "fuzzy" appearance.

DISTANCE FROM EARTH 7,000 light-years.

DIRECTIONS M4 lies just adjacent to Antares, the brightest star in Scorpio. It is closer to the head of the scorpion, that is, just adjacent to Antares on the line between Antares and Delta Scorpii.

SIGNIFICANCE Messier 4 is an excellent object for the beginning astronomer. It appears bright and fluffy, even through small telescopes. As you ramp up the magnification, it is easy to pick out the individual stars within the cluster. There are about 10,000 stars in the cluster, but, of course, it is impossible to pick out all of them from your backyard. It may be one of the closest star clusters to Earth, which accounts for its relative ease of viewing.

M4 is home to one of the most bizarre solar systems in the galaxy—a Jupiter-like planet orbiting a **neutron star**. While not visible with a home telescope, neutron stars are agglomerations of neutrons that weigh as much as many suns but are only a few miles wide. They are as densely packed as the nuclei of atoms, and generally spin very fast. Usually they are the result of massive explosion of conventional stars. How a planet came to orbit a neutron star is not fully understood, but it could have been captured from a neighboring solar system that passed in the vicinity.

Messier 4 is famous for being one of the first objects in the sky that telescopes revealed to be multiple stars. In the late seventeenth and early eighteenth centuries, several astronomers noted how it appeared as one star to the naked eye and then became many upon closer inspection. The same magic trick can still be performed today, of course, and even a good pair of binoculars will serve to resolve this as many objects. Indeed, it is easy to see M4 as multiple stars, but can you see the distinct line-shaped pattern of stars within? This is a challenge for amateur astronomers.

ORIGIN STORY The rather drab name of Messier 4 is a reference to the first cataloguer of many nebulae. French astronomer Charles Messier created a catalogue of objects that vaguely resembled comets but were definitively not, so as to prevent confusion for comet hunters.

POP CULTURE There are no known pop culture references to Messier 4, though its parent constellation, the zodiacal constellation Scorpio, has a long and complex mythos by virtue of being one of the 12 astrological signs.

EAGLE NEBULA

34

PLEIADES

VISIBILITY ● ● ● ● ●

SEASON Fall, but best in November.

QUALITY To the naked eye, a tightly packed cluster of stars, with six stars easily visible, bright, and separable.

DISTANCE FROM EARTH 420 light-years.

DIRECTIONS The Pleiades are next to the constellation Taurus. To find them easily, start from Orion and follow his belt toward Taurus. Then, once you find Taurus, keep going past Aldebaran toward Perseus.

SIGNIFICANCE Even if you live in the brightest city on the planet, you've almost certainly seen the Pleiades. Their image is famous as the hood ornament of automaker Subaru's vehicles, depicting the constellation that is also known as the Seven Sisters. (Subaru is the Japanese name for this cluster.) This is one of the most famous star clusters in history and was depicted on a bronze disc, called the Nebra Sky Disc, which dates back to 1600 BC.

It wasn't until Galileo looked at the Pleiades through his telescope that astronomers realized the Seven Sisters were greater in number than just seven. In the seventeenth century, Galileo saw 36 stars through a very weak telescope. In the late eighteenth century, French astronomer Edme-Sébastien Jeaurat counted 64. How many can you see? Once you identify the cluster, see if you can find and recognize the brightest named stars in the cluster (many of the dim ones don't even have names). These are, in order of brightest to dimmest, Alcyone, Atlas, Electra, Maia, Merope, Taygeta, Pleione, Celaeno, and Asterope. The latter three are generally not visible with the naked eye. All are named for Greek mythological figures.

November is the best month for the Pleiades. During that month these stars can be seen all night long. It is said that Aldebaran looks as though it is chasing the cluster, as it rises immediately before Aldebaran does.

The Pleaides are a great target for unaided astronomy. The number of stars you can see in the cluster is related to how dark the sky is and how adjusted your eyes are to the darkness. If you observe the Pleiades immediately after going outside, and then again after being in the dark for 30 minutes, you'll find that the number of visible stars increases. This is true for many celestial bodies, but it's more dramatic with the Pleiades.

This young cluster, only 100 million years old, is known to have a mass about 800 times that of our Sun. Because of the high velocities of stars within the cluster, they will disperse over the next quarter-billion years, and the Seven Sisters will go off on their own, as all siblings eventually do.

ORIGIN STORY The name *Pleiades* originated with the ancient Greeks, like many major stars and constellations. The Pleiades are also called the Seven Sisters, as the seven most visible stars are said to be seven divine sisters that were born of Pleione and Atlas.

POP CULTURE The Pleiades are one of the best-known clusters in the sky. As noted earlier, the Subaru logo is a visual depiction of these stars. They are mentioned in the Bible and in Homer's *Iliad*. A group of Greek poets in the third century BC called themselves the Pleiade after the cluster. Later, in the sixteenth century AD, a group of French poets took the name, too (La Pléiade).

35

ORION NEBULA

VISIBILITY ● ● ● ● ◐

SEASON Fall and winter.

QUALITY Appears as a bluish star with a red haze around it, distinct in dark regions.

DISTANCE FROM EARTH 1,500 light-years.

DIRECTIONS Find Orion in the sky. Under Orion's belt sits a three-star sword hanging off his side; the middle star is the Orion Nebula.

SIGNIFICANCE The sword hanging off Orion, the hunter, harbors a secret: The middle star of the sword is not really a star at all but, rather, a nebula 25 light-years across, in which thousands of stars are born and die. This is a huge and dense nebula full of forming stars, contained in a massive cloud of particles.

The Orion Nebula, by virtue of its visibility and brightness, has many claims to fame. In 1880, it was the first nebula ever to be photographed. It was sketched by both Charles Messier and Dutch astronomer Christiaan Huygens. It was also one of the first targets of the orbiting Hubble telescope, which imaged it in detail in 1993.

Some astronomers have described the Orion Nebula as a star surrounded by a haze. The haze is the effect from the gas and dust encasing the nebula, the source of the great star factory that defines nebulae. Within the Orion Nebula sit four young, bright stars, collectively known as the Trapezium. The Trapezium is readily visible even with a weak telescope.

There is much to see in the Orion Nebula, its limits determined only by darkness and your telescope's power. Previous astronomers have commented on the nebula's green tinge. Can you see it? The warm, green glow comes from the presence of oxygen in its gaseous clouds. Interestingly, the color of this nebula seems to vary depending on one's eyesight, not the level of telescope magnification.

The human eye has varying levels of color cones—the receptors that detect green, blue, and red light—and those with fewer green cones may have trouble seeing the greenness, except perhaps with a very large scope. Others will be able to see it even with a small scope. Compare notes with your stargazing friends about what they see. With large telescopes, you should also be able to see familiar reddish-tinged areas of the nebula. These are the ionized hydrogen regions that glow from the starlight in the vicinity.

Orion's colors may appear more distinct at higher elevations, as much of its light is filtered by our atmosphere.

ORIGIN STORY The Orion Nebula is named for its parent constellation, Orion, a hunter in Greek mythology who appeared in Homer's *Odyssey*.

POP CULTURE This nebula is named after its associated constellation, and it is just one of many other things called Orion. Other Orions include a line of telescopes, a series of NASA spacecraft, and a book publisher.

36

GHOST OF JUPITER

VISIBILITY ● ● ● ● ◐

SEASON Winter.

QUALITY An orb with a point of light at the center. Through large telescopes, it may look like an eye with a cornea and pupil.

DISTANCE FROM EARTH 1,400 light-years.

DIRECTIONS The Ghost of Jupiter is in the Hydra constellation. Use the bright star Regulus in Leo to find it. The center of Hydra forms a zigzag W shape. Just under Mu Hydrae sits the Ghost of Jupiter.

SIGNIFICANCE When it comes to naming celestial bodies, there are a lot of animals, mythological figures, and spooky or demonic characters. The Ghost of Jupiter, like the Demon Star Algol (see page 80), got its spooky name undeservedly. Other than its pale appearance, there is nothing particularly haunted about the Ghost of Jupiter. It just looks vaguely like Jupiter in a weak telescope, a feature that influenced its name.

The Ghost of Jupiter is a planetary nebula and also one of the easiest to spot and study. To most observers with binoculars or basic telescopes, it appears as a faint circle with a brighter point in the center.

As you improve your skills with a telescope, the Ghost of Jupiter is a good celestial body to return to. If you master basic viewing, the addition of a nebula filter can improve the appearance of the Ghost of Jupiter. Try to see it not as a single circle but rather as a cerulean sphere with an inner oval-shaped orb and a bright star in the very center. It may be difficult to see the star as bluish, but even in urban areas a long exposure may be all it takes. The Ghost of Jupiter is bright enough to stand up for itself among the light-polluted skies of most major cities. It is the remnant of a dying star that has long since exploded. Over the coming millennia the shell will continue to expand and become even more ghostly in appearance.

ORIGIN STORY The Ghost of Jupiter was discovered in 1785 by William Herschel. Through old telescopes, it was said to look vaguely like Jupiter, hence the name.

POP CULTURE The poetic-sounding name has inspired a number of songs. Boston funk band Lettuce has a song entitled "Ghost of Jupiter," and there is even a rock band called Ghosts of Jupiter. English Romantic poet Percy Byssehe Shelley mentions the phrase in his 1820 drama *Prometheus Unbound,* though it is unclear if this is coincidence or intentional.

37

RING NEBULA

VISIBILITY ● ● ● ● ●

SEASON Summer.

QUALITY A bright ring, black in the middle, visible through a larger telescope. With lower-power telescopes, it appears as a hazy oval.

DISTANCE FROM EARTH 2,300 light-years.

DIRECTIONS Find the constellation Lyra, which is easy to spot because of bright star Vega (see page 72). On the opposite end of the constellation from Vega sit stars Sheliak and Sulafat. Two-thirds of the way down the imaginary line from Sulafat to Sheliak sits the Ring Nebula.

SIGNIFICANCE Thousands of years ago, before the advent of the telescope, virtually everything in the sky (beside the Sun and Moon) looked like a point of light. Ancient astronomers thought that the sky was a black veil with white light behind it, and stars were merely holes poked in the veil. It was sacrilege to think that another body in the sky might not be merely a point of light, and Galileo was excommunicated from the Catholic Church for his observations that the night sky was more complicated and rich than previously thought.

Imagine, then, how strange the Ring Nebula must have appeared when it was finally seen through a telescope—a fluorescent hoop, lonely and perfect in the sky. As this was only the second nebula ever discovered, it must have been quite the novelty. What could have caused such a perfect-looking hoop to form? The answer, as you probably suspect, was a giant explosion. The exploding star blew off its outer shell. The ring that gives the nebula its name is the shell of that star slowly moving outward into the depths of space.

At the center of the Ring Nebula sits a tiny white dwarf star, the core remnant of the original star that exploded to produce the nebula. This probably won't be visible no matter how large your telescope is, though with a long-exposure photograph you may be able see it directly in the center of the ring.

With smaller telescopes, the appearance of the nebula is often compared to a bagel or donut. Through larger telescopes, or with a long exposure, the greenish-reddish tinge to the bagel may be more apparent. Also, with a larger telescope you may be able to see that the nebula is actually not a circle, but more like an oval, slightly stretched and redder at its ends.

ORIGIN STORY This nebula is so named for its appearance as a ring. It was discovered in 1779 by French astronomer Antoine Darquier.

POP CULTURE A US Postal Service 33-cent stamp featuring the Ring Nebula as imaged by the Hubble Space Telescope was released in the year 2000. American travel writer John Lawson Stoddard (1850–1931) wrote a poem to the nebula, entitled "To the 'Ring Nebula.'"

EAGLE NEBULA

VISIBILITY ● ● ● ● ●

SEASON Summer to fall.

QUALITY A fuzzy star as seen with the unaided eye; a fuzzy red patch with points of light as seen with a telescope.

DISTANCE FROM EARTH 7,000 light-years.

DIRECTIONS The Eagle Nebula is near the constellation Scutum, but it is easier to find when you start from Sagittarius. Imagine a line between Kaus Australis and Kaus Media (on the spout of the "teapot"), and follow this line up until you are almost parallel to the bottom star in Scutum. The Eagle Nebula is just barely beyond this point. M17 (see page 101) and M18 precede it and are on the same line.

SIGNIFICANCE We return to that star-rich region of the sky by Sagittarius, dense because of its proximity to the plane of the Milky

Way. The Eagle Nebula is one of the best-known nebulae in the region, made famous by *The Pillars of Creation,* a famous Hubble Space Telescope photograph of a 9-light-year-long triad of dusty pillars floating in its immense field of stars. These formations show the beginnings of the creation of protostars, the stage before nebula gas accumulates into real stars. The three pillars are actually visible from your backyard with a telescope. However, this requires excellent conditions and a powerful telescope at high magnification.

The Eagle Nebula, as a stellar crèche, is particularly dramatic. It has a vivid red glow, the by-product of a large agglomeration of hydrogen—fuel for new stars. There are an estimated 400 stars milling around the nebula, which have accumulated over millions of years from its gaseous depths.

Depending on the size of your telescope, the Eagle Nebula can vary in appearance. Its distinct red color is apparent even with small instruments. With larger instruments, the striated lines of gas become visible, and you can see more and tinier stars within. Try looking at it at different times of night and in different seasons. Can you see the shape of the nebula as a bird of prey? It may not be obvious at first, but keep trying.

The Eagle Nebula is home to a few young, very unstable large stars that may explode dramatically in the coming million years, including one star weighing in at 80 times the mass of our Sun. Stars of such mass are rare, mainly because their lifetimes are so short—they are almost certainly doomed to explode in dramatic supernova blasts, the most powerful and brightest explosions in the universe. If you are lucky, perhaps you will witness one in the Eagle Nebula.

ORIGIN STORY This nebula is so called because its shape is said to resemble an eagle's head.

POP CULTURE The Eagle Nebula is a favorite of science fiction writers and has appeared in *Star Trek: Voyager, Babylon 5,* and briefly in the film *Contact* (adapted from a book by Carl Sagan).

LAGOON NEBULA

M8

VISIBILITY ● ● ● ● ●

SEASON Summer and fall.

QUALITY Oval-shaped bright patch, with a bow tie–shaped nucleus visible as magnification increases.

DISTANCE FROM EARTH 4,100 light-years.

DIRECTIONS Find the familiar "teapot" of Sagittarius. On the side of the teapot facing Scorpio—the spout side rather than the handle side—locate the scalene triangle formed by the three stars of the spout. If you imagine a line from Kaus Australis (the star at the teapot's base) directly between the other two stars in the triangle, it will touch both the Lagoon Nebula first and then the Trifid Nebula (see page 105).

The distance from Kaus Australis to the two nebulae is about half the distance from the bottom to the top of the triangle.

SIGNIFICANCE The Lagoon Nebula is one of the most visible nebulae in the night sky. You can even see it with the naked eye, though it won't look like much without a telescope. Yet even with a basic pair of binoculars the Lagoon Nebula will start to come into focus like a bright splash of paint on a dark sky. Some believe that the nucleus, or center of the nebula, resembles a bow tie or an hourglass. Can you see it?

The nebula is reddish if viewed without the interfering atmosphere. However, the only way you can really see it as red from home is by taking a longer-exposure photograph with a camera or camera attachment on your telescope. Because of its situation in the sky, viewing the Lagoon Nebula is actually easier in more southerly parts of the Northern Hemisphere, where it appears more vivid than in, say, northern Canada or Alaska.

The most impressive aspect of the Lagoon Nebula is its size. Through a larger home telescope, you'll find that you can scan around the Lagoon Nebula for a long time without seeing all its sights. It is very rich and large—over 100 light-years across—with a small star cluster visible within its depths. It is actively expanding, too, as gas coalesces into dense new stars. The previously mentioned bow tie–shaped inner nebula may be harder to see, depending on your telescope's size.

The Lagoon Nebula is neighbor to the Trifid Nebula (see page 105), though slightly less colorful and slightly larger in span. Both are good companions to view together on a dark night.

ORIGIN STORY Unusually, this nebula's name does not originate in mythology. Rather, it is a reference to the appearance of the central part of the nebula as two separated lagoons. It was first recorded in 1654 by Sicilian astronomer Giovanni Battista Hodierna.

POP CULTURE The Lagoon Nebula's biggest claim to pop culture fame is its appearance in the 2002 Disney movie *Treasure Planet*, where it formed a major setting for the story. Jerry Sohl's science fiction novel *The Spun-Sugar Hole* (1971) refers to the Lagoon Nebula.

IRIS NEBULA

40

CRAB NEBULA

VISIBILITY ● ● ● ● ●

SEASON Fall through spring.

QUALITY When viewed through binoculars, this appears as a very faint small haze, if you can see it at all. (Try looking slightly to the side of the nebula.) Through a telescope it appears as an oblong nebula that is brighter along one axis.

DISTANCE FROM EARTH 6,500 light-years.

DIRECTIONS Start from the Taurus constellation. The portion of Taurus that looks like a tuning fork has a star toward the end called Zeta Pegasi. On the inside of the tuning fork, just off Zeta Pegasi and toward the center of the constellation, sits the Crab Nebula.

SIGNIFICANCE The Crab Nebula may be the most famous supernova in recorded history. In the year 1054 a massive supernova explosion emanated from the Crab Nebula (which, prior to then, was a star). The exploding supernova was temporarily the third-brightest object in the sky, after the Sun and Moon, and was visible during the daytime for almost a month. Japanese astronomers recorded it as brighter than

Jupiter and called it a "guest star." They were right about its guest status, as it faded after a few weeks yet remained bright in the sky for two years. The appearance of a new star in the sky was seen as foreboding to most of the Arab, Chinese, and Japanese astronomers who noticed it. However, in the midst of the Dark Ages, most of Europe seemed to have hardly noticed—save for a few in Switzerland where notes of the star appear in a journal. What is called the Crab Nebula today is the remnant of this massive explosion.

Because you need at least a 3-inch telescope to view the Crab Nebula, it ranks as only a 2 on the visibility scale. However, it is a very notable and important nebula, even if it's not the brightest. Astronomers have used the Crab Nebula to experimentally verify many aspects of astronomy. For instance, when it passed behind Saturn's moon Titan, observations of how the X-rays from the Crab Nebula were distorted provided evidence of the composition of Titan's atmosphere.

The Crab Nebula's visible details vary with conditions. When viewed through a large telescope, the crustacean shape will be more apparent. The core should appear brighter than the remainder of the nebula. Little stars may appear in the background of the nebula, shining through it. Under certain conditions you may be able to see the red streaks mixed with the green glow of the center portion. Some say it more resembles a spider than a crustacean. Can you see it as a crab?

ORIGIN STORY In 1840, English astronomer John Bevis observed the Crab Nebula with a large telescope and decided that it looked like a crab, hence the name. It is also known as M1 (Messier object 1) because it was the first object that Charles Messier mistook for a comet; its discovery led to him writing his famous catalogue.

POP CULTURE The 2007 video game *Mass Effect* featured imagery from the Crab Nebula. Poet Ann Shaffer has a poem entitled "Under the Crab Nebula." George R. R. Martin's 1986 novel *Tuf Voyaging* includes a wry reference to "the Empress of the Crab Nebula." William S. Burroughs's *Nova Express* features a chapter on the nebula.

41

OWL NEBULA

VISIBILITY ● ● ● ● ●

SEASON Winter and spring.

QUALITY A circular blue puff; through large telescopes, you'll see three stars in the shape of a triangle appear at the center of the circle.

DISTANCE FROM EARTH 2,600 light-years.

DIRECTIONS The Owl Nebula is located between two stars of the Big Dipper. Find the ladle of the dipper and locate the two stars on the bottom—the long side that doesn't touch the handle. These are Beta Ursae Majoris and Gamma Ursae Majoris. Imagine a line between the two. The Owl Nebula is three-quarters of the way from Gamma to Beta. Find this with binoculars first before switching to a telescope.

SIGNIFICANCE Nebulae are basically clouds of dust and gas, so it's fitting that people often gaze at nebulae and see shapes in the same manner as clouds. The Owl Nebula is an especially creative example. When English astronomer William Parsons saw this nebula in the nineteenth century, he thought the faint inner dark circles, roughly symmetrical within the larger pink circle, resembled a barn owl.

Depending on the season, this could also appear upside down, which might make it look less like an owl and more like a pig's snout.

There is a triad of stars toward the center of the nebula, but these are very faint, as they are white dwarfs. Only the largest-lensed telescopes will be able to make them out. The ghostly blue circle of the nebula has traces of red surrounding it, too, which may be visible in good conditions or with the right equipment.

The Owl Nebula, by virtue of being a light-year in diameter, is quite large in the night sky—though its brightness is not great. This may be because of the nebula's diffuse nature: The expanding sphere of gas has a mass about one-tenth of that of our own Sun. This is too small to create any new stars, and its gas will eventually become invisible and sparse as it expands into space. As with most nebulae, it is spherical as the result of a large stellar explosion, which remains in the tiny white dwarf at the center of the nebula.

Though dim under a light-polluted sky, the Owl Nebula's fame and beauty makes it a worthy member of any amateur astronomer's list of must-see objects.

ORIGIN STORY Though discovered in the late eighteenth century by French astronomer Pierre Méchain, this nebula gets its name from an owl-like sketch drawn by Englishman William Parsons in 1848.

POP CULTURE There are no known pop culture references to the Owl Nebula.

42

VIRGO CLUSTER

VISIBILITY ● ● ● ● ●

SEASON Summer.

QUALITY A series of galaxies in a wavy line, smeared across the sky near Virgo.

DISTANCE FROM EARTH 65 million light-years.

DIRECTIONS The galaxies of the Virgo Cluster are spread between Denebola (in the constellation Leo) and Vindemiatrix (in the constellation Virgo). If you can locate these two stars in their respective constellations, scan the skies in a line between them. You won't have to look far before you start to make out the galaxies.

SIGNIFICANCE This marks our first foray into viewing objects outside of our own galaxy. The Virgo Cluster is a long stretch of clustered galaxies. There is much to see in this region, which extends some

8 degrees across the night sky. Its length in the night sky means all of it cannot possibly be seen in one viewing.

Galaxy clusters are the largest collective bodies that exist in the universe. Beyond these, there is no larger object to observe. In previous selections, you have observed clusters that contain thousands or tens of thousands of stars. Yet galaxies, ours included, contain between 40 billion and 1 trillion stars. So a cluster of galaxies can contain 1,000 trillion stars or more.

Look closely at each of these galaxies. Each is teeming with stars and planets. In your field of view, you could be looking at other planets whose inhabitants are looking up at the sky with their telescopes.

The best way to approach viewing the Virgo Cluster is to start from the west and move east. It may be easier to follow the trail of galaxies this way—the longest arc of continuous galaxies is known as Markarian's Chain and is not too hard to see. Once you've mastered Markarian's Chain, feel free to veer up and down slightly and expand the scope of your vision.

There are two very different kinds of galaxies visible in the Virgo Cluster: spirals and ellipticals. Spiral galaxies, like our own Milky Way, resemble the classic image of a galaxy—a swirled shape with a central bright point. However, there are actually multiple kinds of spiral galaxies, including barred spirals, which simply means that the lines emanating from the nucleus jut out in straight lines before curving.

The second kind of galaxy, elliptical, is more mysterious. It appears as bright points with a blur around it of decreasing brightness, sort of like a lens flare. Elliptical galaxies tend to be older than spiral galaxies and slightly yellower. If you study the elliptical galaxies in the Virgo cluster closely, you should notice how different from spirals they appear.

The spiral galaxies will almost certainly be more interesting than the elliptical ones. With increasing magnification, you will be able to see more detail in the spiral galaxies—even their individual arms with larger telescopes and possibly some of the larger nebulae. You may also be able to discern which are barred and which are not.

ORIGIN STORY The Virgo Cluster of galaxies was named after its proximity to the constellation Virgo. Most of the galaxies were discovered in the late eighteenth century as optics technology was improving.

POP CULTURE A supercomputer at the Indian Institute of Technology is named for the Virgo Supercluster, the larger group of gravity-linked galaxies that includes the Milky Way. Drum and bass artist Smooth has an album titled *Virgo Cluster*.

BODE'S GALAXY

M81

VISIBILITY ● ● ● ● ●

SEASON Winter and spring.

QUALITY A brilliant spiral galaxy; as viewed from above, it appears oval shaped. A second galaxy, M82, may appear next to it. M82 is smaller and looks like a star with a tail coming off either end.

DISTANCE FROM EARTH 11.8 million light-years.

DIRECTIONS Start from the Big Dipper. Imagine a line between the two diagonal stars in the rectangular "ladle" portion, Phecda and Dubhe. If you continue past Dubhe a little less than one ladle's width, you'll chance upon M82 first, then M81 adjacent to it (they usually appear in the same view in your eyepiece).

SIGNIFICANCE For an astronomer, living in the Northern Hemisphere can seem like a drag. The bright center of the galaxy is visible only from the Southern Hemisphere, as are the two satellite galaxies of the Milky Way, the Magellanic Clouds. With just one-tenth of the population of the Northern Hemisphere, the Southern Hemisphere also has far less light and atmospheric pollution. Furthermore, many of the brightest and nearest stars are more prominent from the Southern Hemisphere.

M81 is a notable exception. It is visible only from the Northern Hemisphere. And lucky for us, since it is one of the brightest galaxies in the sky, it can be viewed through a simple pair of binoculars. It is also rumored to be visible to the naked eye in the darkest of nights and under the clearest of skies.

One of the most prominent Messier objects included in the French astronomer's catalogue, M81 is part of the Local Group of galaxies, a cluster of gravitationally bound galaxies of which the Milky Way is a member. And what a brilliant galaxy it is! Its hundreds of billions of stars increase in vividness as your magnification increases.

As you might have noticed, M81 has a bonus feature: It is immediately adjacent to M82, which appears through your telescope as a smaller galaxy oriented edge-on. As such, unfortunately you can't see M82's spirals; instead, it looks like a long splat of stars with a bright central point. (The central point of the galaxy is the bright cluster surrounding a **black hole**.) M82 and M81 are each about the same distance from Earth.

You may be able to see a faint glow from the bright center of the M82 galaxy, depending on conditions. These hydrogen gas emissions are not completely understood by scientists but could be the result of a nebula interacting with the central black hole in the galaxy.

In the early 1990s, a supernova explosion in M81 temporarily brightened the galaxy (the supernova appeared as a huge, bright star within). As in any other visible galaxy, this could certainly happen again at any moment, so a supernova alert is issued for both M81 and M82.

ORIGIN STORY M81 and M82 were both Messier objects, listed in French astronomer Charles Messier's catalogue of celestial objects commonly confused with comets.

POP CULTURE M81 has made several appearances in *Star Trek*'s various series. The image of the two galaxies together is one of the most popular astronomical images and is frequently reproduced.

TRIANGULUM GALAXY

44

ANDROMEDA GALAXY

VISIBILITY ● ● ● ● ●

SEASON Fall, winter, and spring.

QUALITY A blue sphere with a halo around it in an oval; two smaller satellite galaxies are sometimes visible, forming a 120-degree angle relative to the center of the galaxy.

DISTANCE FROM EARTH 2.5 million light-years.

DIRECTIONS Find the Andromeda constellation (next to Cassiopeia). Beta Andromedae hangs off one side of Andromeda, heading out from Alpheratz. Follow Beta Andromedae up to Mu Andromedae. Then continue in the same direction another Beta–Mu distance to find the Andromeda Galaxy. Start with binoculars if you get lost.

SIGNIFICANCE Currently the Andromeda Galaxy is the closest galaxy to our own, the Milky Way. That won't always be the case, though. In about four billion years, the two galaxies will merge in a giant cosmic collision and, well before that, the Andromeda Galaxy will take up so much of the night sky that our nights will no longer be dark. It is unknown whether Earth will be much affected by the transition (provided it is still around then), but, since space is so vast, it is possible that our solar system could pass through unscathed. Because the Andromeda Galaxy is so much larger than the Milky Way—with more than twice as many stars—our cosmic address may have to change to reflect this new home.

For now, though, the Andromeda Galaxy is another point of light to the unaided eye, and one of only a few galaxies visible without a telescope. With a weak telescope, you should be able to make out at least one of the two satellite galaxies that orbit Andromeda—M32 and M110. M110 appears farther from the core and runs approximately perpendicular to the plane of the halo. It is easier to see than M32, which is closer to the Andromeda Galaxy and appears on the opposite side. Together the three galaxies form about a 120-degree angle with each other. Andromeda has a huge number of satellite galaxies, at least a dozen, not all of which are visible. (The Milky Way has two as well, yet they are visible only from the Southern Hemisphere.)

Andromeda is a great object to use as a test for filters. If you are at the skill level of using filters over your lenses, experiment with one on Andromeda to see whether you can spot Andromeda's spiral arms, the bluish-purplish haze that runs through the halo, or the "**dust lanes**"— the dark spots that run like veins through its arms.

ORIGIN STORY The Andromeda Galaxy is also known as the Great Spiral Galaxy. Ancient astronomers noted its cloud-like appearance, and it is named after its home constellation, Andromeda, which is named after the daughter of Cassiopeia in Greek mythology.

POP CULTURE Superman's homeworld Krypton, in at least one comic, has been described as being in the Andromeda Galaxy. Many episodes of *Doctor Who* refer to aliens from this galaxy. Colin Kapp's novel *The Patterns of Chaos* takes place partially in the galaxy. Poet Brenda Shaughnessy has a collection entitled *Our Andromeda*.

TRIANGULUM GALAXY

VISIBILITY ● ● ● ● ●

SEASON Fall.

QUALITY A bright galaxy, which may appear brighter by averting one's vision.

DISTANCE FROM EARTH 3 million light-years.

DIRECTIONS The Triangulum Galaxy appears halfway between the last star of the Pisces constellation and the acute vertex of the Triangulum constellation, Alpha Trianguli.

SIGNIFICANCE The Triangulum Galaxy may be the most distant celestial body that one can see with the naked eye. It is also one of the largest in terms of mass and volume—a stunning spiral galaxy spanning 60,000 light-years across and including 40 billion stars. The Triangulum may actually be a satellite galaxy of behemoth

Andromeda (see page 135), slowly orbiting its parent galaxy. Perhaps when Andromeda and the Milky Way merge, Triangulum will then orbit the new supergalaxy that they combine to create.

Depending on local conditions, the Triangulum Galaxy will be visible either directly or indirectly with the naked eye. In urban areas, you will almost certainly have to use averted vision to see it. Test first with a low-magnification telescope (or binoculars) before moving to a high-powered telescope.

Once you have it in your view, the next challenge is to find and count Triangulum's arms. From our perspective they appear as a backward S shape as they snake out from the bright center of the galaxy. Under certain conditions you may be able to see the veinlike dust lanes around some of the arms.

The Triangulum Galaxy is full of open clusters and nebulae that are visible from Earth. There are literally dozens of these and you probably can see at least several of them even with a small scope. Playing with filters or viewing this galaxy under darker conditions can reveal all these numerous features.

Many of the features of Triangulum can be captured in long-exposure photographs, most notably the bright colors of its internal nebulae. If you have the capability of taking a 30-second exposure or longer, you may be surprised at how many details you can see in the galaxy.

ORIGIN STORY Charles Messier included the Triangulum Galaxy in his catalogue of stars, but previous astronomers probably saw it without being aware of its significance.

POP CULTURE In "Where No One Has Gone Before," an episode of *Star Trek: The Next Generation*, the crew accidentally travels to the galaxy. There have been hints, too, that the main character of the 1982 movie *E.T. the Extra-Terrestrial* was from this galaxy.

46
WHIRLPOOL GALAXY

VISIBILITY ● ● ● ● ●

SEASON Winter and spring.

QUALITY A perfect spiral galaxy with a bright white spot at the end of one arm (an elliptical galaxy companion).

DISTANCE FROM EARTH 25 million light-years.

DIRECTIONS Start from the Big Dipper. The Whirlpool Galaxy is just above the last star in the Big Dipper's handle. Megrez and Dubhe, two stars in the Big Dipper's ladle, point very close to it.

SIGNIFICANCE The last galaxy in our list is also the grandest. The Whirlpool Galaxy resembles a perfect spiral, yet with a strange, bright interloper off the edge of one its arms, which looks like a globe-shaped light bulb. This interloper is NGC 5195, an irregular dwarf galaxy that is gravitationally interacting with the larger Whirlpool.

For being 25 million light-years away from Earth—about 200 times as far as the Milky Way is wide—it is incredible that the Whirlpool can be seen with binoculars. It is actually slightly smaller than the Milky Way in terms of size, and its mass is about one-third of our galaxy's.

The companion galaxy, NGC 5195, is a strange dwarf galaxy with no obvious discernible shape. It is called an irregular galaxy, one that doesn't readily fit into the three categories of galaxies that have been observed. This is likely because it has been stretched and affected by larger galaxies, like the Whirlpool, to the point that its original form may no longer be discernible. The tendril-like connection between the arm of the Whirlpool and NGC 5195 has been described as having a "dust bridge"—that is, a nebula-like vein near where they connect. You may actually be able to make out some of the nebula-like features of the galaxies under certain conditions. The Whirlpool Galaxy is known for being sensitive to light pollution—try to look at it during darkest times of night, or in darker regions, for best viewing.

The Whirlpool often looks better with smaller telescopes, through which it is easier to view the bright core of both galaxies along with some of the arms. Through larger telescopes you will be able to see more variation in the arms and probably pick out larger bright nebulae within the Whirlpool. The arms appear to wrap around the core about three times as they move out in a spiral. Can you discern the full three revolutions? There are dust lanes along the inner spiral that may be visible with certain filters or longer exposures.

ORIGIN STORY The Whirlpool Galaxy was one of the first galaxies discovered by seventeenth-century astronomers who, at the time, called them "spiral nebulae," not knowing what they were. It was given the nickname Whirlpool after its whorl-like appearance.

POP CULTURE The computer game series *Homeworld* featured this galaxy. An iconic Hubble photograph of the galaxy is commonly reproduced for mugs, shirts, and computer backgrounds.

47

COMETS

VISIBILITY Varies seasonally.

SEASON Varies. At any given moment, there tends to be at least one or two comets visible with a telescope, and about one a year visible, if faintly, to the naked eye.

QUALITY Looks like a bright hazy star in the sky, usually followed by a long tail.

DISTANCE FROM EARTH Varies. Only comets within our solar system are visible, so usually within 100 million miles or less.

DIRECTIONS Varies, and can change rapidly. Check online before viewing.

SIGNIFICANCE Galactic clusters like the Virgo Cluster are the most distant and largest objects you've seen with your telescope. And comets are by far the smallest. Many are only a few miles in diameter. So how are they visible, sometimes even with the naked eye? It turns out it's not the size that counts, but the luminosity: Comets are extremely bright in the sky, by virtue of the fact that they are, more or less, made of snow. This is more intuitive than you might think. Consider how bright and reflective a snowy mountain is, such that most skiers and snowboarders wear sunglasses or goggles even on cloudy days.

Future comet appearances are harder to predict with certainty, as they often are not discovered until a year or two before they become visible to amateur astronomers. There are a few comets that return regularly, such as Halley's comet, which appears about once every 75 years (see the table following). Others, such as the comet Hale-Bopp, appear every 2,500 years—meaning that when it approached Earth in 1997, it was considered a "new" comet (though there may be evidence that the ancient Egyptians recorded its passing. Still other comets pass once and fall into the Sun, never to return. It is thought that although comets shed mass when they are close to the Sun, they gain mass in the intervening years when they return to the distant Oort Cloud, a sparse field of ice and dust beyond Pluto.

It is impossible to give a comprehensive list of comets because most of them have not yet been discovered. However, the list that follows provides some information on the known upcoming comets and when they will be arriving in Earth's view. Check online for more details on precise locations—comets move much faster through the sky than other objects, even most planets.

Comets are notoriously finicky. Halley's comet, which has a reliable 75-year orbit around the Sun, varies greatly in brightness. While it will return to swing around the Sun in 2062, it is not known how bright it will be.

ORIGIN STORY The word *comet* comes from a Greek word meaning "long-haired star." It is related to the Greek word for hair.

POP CULTURE Fictional depictions of comets include the 1998 disaster film *Deep Impact*, Edgar Allan Poe's "The Conversation of Eiros and Charmion," George R. R. Martin's *A Clash of Kings*, H. G. Wells's *In the Days of the Comet*, Walt Whitman's "Year of Meteors," and Adrienne Rich's "Planetarium."

COMETS VISIBLE FROM THE NORTHERN HEMISPHERE AT NIGHT

COMET NAME	DATES OF VISIBILITY
Comet C/2013 US10	late 2015–summer 2016
P/2010 V1	winter 2015–summer 2016
67P/Churyumov-Gerasimenko	fall 2015–winter 2015
C/2013 X1	spring 2016–fall 2016
45P/Honda-Mrkos-Pajdušáková	winter 2016–spring 2017
2P/Encke	spring 2017
41P/Tuttle-Giacobini-Kresak	spring–early summer 2017
21P/Giacobini-Zinner	late summer 2017
46P/Wirtanen	fall 2017–winter 2018
Halley's comet	2062 AD
comet Hale-Bopp	4385 AD

48

THE SUN

[
WARNING: YOU NEED A SOLAR FILTER FOR THIS.
Do not use unfiltered telescopes, binoculars, or even your naked eye to look directly at the Sun, as the intense light can permanently damage your eyes. A solar filter is an inexpensive device that fits over the lens end of your telescope and blocks out almost all the dangerous light. They tend to cost between $20 and $50. With one of these you can view the Sun directly and also see its sunspots and corona.
]

VISIBILITY ● ● ● ● ●

SEASON All.

QUALITY A bright orange ball with wispy emanations from the corona and with dark surface sunspots.

DISTANCE FROM EARTH 150 million kilometers (9 light minutes— the distance an unimpeded ray of light travels in one minute).

DIRECTIONS Fit the solar filter onto the end of your telescope, making absolutely certain that it is screwed in tightly. Test first by aiming the telescope toward the sky to ensure that your view is totally black. Then, point it toward the Sun, wearing sunglasses as you do so. Note that there are also special-purpose solar telescopes made specifically for solar observing.

SIGNIFICANCE If you cannot wait until night, or if you live in a particularly bright city like Manhattan, you can always do astronomy during the day and look at the Sun instead. The Sun is one of most peculiar and rich objects in the sky, yet because you cannot look at it directly you may not normally think of it as such.

Here's where a solar filter comes in. Solar filters allow you to look closely at the Sun's surface without going blind. Once you do, you'll find the surface of the Sun is not uniformly orange as it seems. Its strong magnetic fields create a lattice of tube-shaped arcs of plasma, about the width of the Earth, which snake in and out of the Sun's dark spots like untethered water hoses.

Simply by virtue of its size and the massive amounts of energy in its depths, there is plenty happening on the Sun's surface. The magnetic field generated by the Sun also causes sunspots, dark bruises on the surface that you can see with your solar filter–outfitted telescope. The sunspots on the surface increase and decrease in number with a very predictable cycle—every 11 years, the number of sunspots increases to a peak. Sunspots appear in certain regions on the surface of the Sun, typically opening above the equator and moving toward it.

Since the Sun revolves about once a month, you can actually watch sunspots move across its surface day by day. Sunspots look vaguely like scabs, with the surface of the Sun looking a bit like skin. You won't need great magnification to study the Sun, as it is quite large in the sky already.

When you observe the Sun and count its sunspots, you're following in the footsteps of astronomers who, for the past 400 years, have made careful observations of the Sun that are still referred to today. For instance, records of sunspot activity dating back to the early 1600s have confirmed a theory that the number of sunspots correlates to a slight increase in temperature on Earth. This is a very subtle effect, but the last Little Ice Age took place during a period of reduced sunspot activity.

With a solar filter you can also view the Sun's corona, the hot gaseous emanations from the surface. These appear around the Sun as an ever-shifting halo and are especially visible during a total solar eclipse (see page 148).

ORIGIN STORY Galileo was the first to observe sunspots, although his method was very dangerous and he did not use a filter. He nearly went blind as a result. However, his observations and discovery of sunspots paved the way for solar astronomy.

POP CULTURE The Sun is one of the most common astronomical symbols that appear in art. The word *sunny* is synonymous with "happy," and most words that begin with the *sol* prefix have their origin in the Sun. This includes *parasol* (an umbrella for repelling the Sun), *solarium* (a room that lets in sunlight), and *solstice* (the time of year when the Sun is at its highest or lowest point).

BODE'S GALAXY

SOLAR ECLIPSE

VISIBILITY ● ● ● ● ● In the rare event that they occur.

SEASON Varies. Eclipse season is said to occur when the Moon is in the cycle of its orbit that brings it closer to Earth (though this is independent of Earth's seasons, which occur because of the Earth's tilted axis).

QUALITY

- **PARTIAL AND ANNULAR** A sudden darkening of the sky, like a cloudy day, as the Moon passes between the Earth and Sun.
- **FULL** The sky descends into darkness and stars come out for several minutes.

DIRECTIONS For more information on upcoming solar eclipses, refer to Appendix A: Schedule of Solar Eclipses (see page 154).

SIGNIFICANCE A solar eclipse occurs when the Moon passes between the Earth and the Sun—that is, the three bodies align and the shadow of the Moon stretches across the Earth's surface and briefly occludes the Sun. As you might have inferred, solar eclipses can occur only during the daytime, when the Moon is completely new.

As the Moon slowly recedes from the Earth, solar eclipses become more and more rare. Billions of years ago the Moon was much closer to the Earth and solar eclipses were common. But since then the Moon has receded at the rate of about one centimeter per year—its gravitational energy absorbed into Earth's tides, creating waves on Earth that helped early life proliferate. (You can thank the Moon for helping life get started—without it, humans might not exist.)

If you've witnessed a solar eclipse, you may know how brief and rare they are. The longest tend to last no longer than a few minutes. Any given point on Earth's surface is likely to experience a total solar eclipse only once in a hundred years. So if you want to see one—especially a total solar eclipse—you will likely have to travel. In a few hundred million years, total solar eclipses will no longer happen at all—a tragedy, but one you won't be around to witness.

There are three kinds of solar eclipses that you can witness from the ground on Earth:

1. **TOTAL SOLAR ECLIPSE** occurs when the Moon completely blots out the Sun, transforming a sunny day to total darkness and revealing, in the absence of clouds, stars that you'd see only under a night sky. The longest total solar eclipses last only a few minutes. As Earth spins, the path of darkness moves quickly across its surface.

   Total solar eclipses also provide a rare opportunity to see, in vivid detail, the Sun's corona—which, with the Moon occluding the Sun's circle, appears as a series of wispy rays emanating from around the eclipse.

DOUBLE-CHECKING EINSTEIN WITH SOLAR ECLIPSES

Some amateur astronomers may be interested in using the rare total solar eclipse to verify Albert Einstein's law of general relativity, which states that light will bend around objects with very high mass. In 1919, during a total eclipse, Sir Arthur Eddington photographed the stars of the Hyades cluster (see page 95) several **arcseconds** (one-sixtieth of an **arcminute**, which is one-sixtieth of a degree) away from where they "should" have been—proving that light from the stars indeed bent around the Sun and providing empirical evidence for Einstein's theory. While it takes a sensitive telescope to be able to measure and calculate position, modern telescope technology has made this experiment possible for hobbyists.

2. **PARTIAL SOLAR ECLIPSE** occurs when the Moon only partially blocks the Sun and appears as a dark disk occluding the corner or edge of the Sun. When there is a total solar eclipse in one region of the world, other regions experience a partial solar eclipse. In the moments before a total solar eclipse reaches **totality**—the point at which the Moon completely covers the Sun—the eclipse is partial.

3. **ANNULAR SOLAR ECLIPSE** occurs when the Moon covers the Sun almost completely, but the Sun still shines around the edges of the Moon, producing a ring of fire on a black disk. Annular solar eclipses occur because the Moon's orbit is eccentric, meaning it is not orbiting in a perfect circle—in fact, it varies by about 40,000 kilometers (25,000 miles) in its distance to Earth as it revolves.

There is a general agreement among eclipse chasers that each of the three kinds of eclipses correlate to different degrees of awe. A partial eclipse is worth up to an hour's trip to witness; an annular eclipse is worth a day's drive; and a total solar eclipse is worth any and all hassle to witness.

Solar eclipses are incredible enough without a telescope, but with a telescope you can see some interesting things. Put a solar filter on your telescope to watch the black disk of the Moon as it starts to occlude the Sun; you'll see the Sun's corona making waves around the Moon. In the moment of totality, you can aim your telescope elsewhere in the sky and, briefly, see stars that would normally be visible only in the opposite season.

POP CULTURE Solar eclipses historically portended doom, gloom, and dark magic. Many wars and battles with grisly outcomes were said to have been doomed by solar eclipses, including the taking of Constantinople by the Turks in 1267. William Shakespeare incorporated eclipses into many of his plays, including *King Lear*, when Edmund says, "And pat he comes like the catastrophe of the old comedy: my cue is villainous melancholy, with a sigh like Tom o' Bedlam. O, these eclipses do portend these divisions!"

50

LUNAR ECLIPSE

VISIBILITY ● ● ● ● ● When they do occur.

SEASON Rare, but more common than solar eclipses. About one lunar eclipse happens each year, though the season varies.

QUALITY The full Moon fades and begins to turn black, then dark red.

DIRECTIONS Refer to the list of upcoming eclipses in the Appendix B: Schedule of Lunar Eclipses (see page 156).

SIGNIFICANCE A lunar eclipse occurs when the Earth passes between the Moon and the Sun—that is, the three bodies align and Earth casts a shadow on our Moon. Lunar eclipses were one of the first instances that proved the world was round: You can see the shape of Earth as its shadow passes over the Moon. As you might have guessed, lunar eclipses can occur only during the nighttime, when the Moon is completely full and the Sun is on the opposite side of the planet.

If you point your telescope at the Moon as the disk of the Earth begins to occlude it, you should be able to make out the Earth's shadow as it moves quickly across the surface. As this happens, sharp topography will appear to jump out across the Moon's surface. When Earth's shadow covers it wholly, the Moon takes on a strange red appearance, an effect of the Sun refracting through the edges of Earth's atmosphere and, from there, shining on the Moon. Similar to the red-orange sky at sunset, this same palette is projected onto the lunar surface from the many horizons on the bright side of the Earth. No wonder that many people find lunar eclipses eerie, even unsettling.

As with solar eclipses, there are three different kinds of lunar eclipses:

1. **TOTAL ECLIPSE** occurs when the Earth's shadow completely swallows the Moon.

2. **PARTIAL ECLIPSE** occurs when the edge of the Earth's shadow shades part of the Moon.

3. **PENUMBRAL ECLIPSE,** which is more subtle and generally less interesting, occurs when the Earth doesn't completely block the Sun from the Moon's perspective but blocks only a little of its light. The effect of a penumbral eclipse is similar to when a bug lands on a light bulb, in that the room gets darker but stays lit. During a penumbral eclipse, the Moon may appear to darken and turn slightly brown.

POP CULTURE Like solar eclipses, lunar eclipses were great portents of the ancient world. Plutarch, an ancient Roman, noted how citizens tended to panic when lunar eclipses occurred. During the third Macedonian War, an astronomer named Gaius Sulpicius Gallus was sent to calm the Roman soldiers by explaining the origin and science behind lunar eclipses.

APPENDIX A
SCHEDULE OF SOLAR ECLIPSES

TYPE & DATE	WHERE VISIBLE
Partial; September 13, 2015	Southern Africa, Antarctic Ocean, Indian Ocean
Total; March 9, 2016	Sumatra, Borneo, parts of the Pacific Ocean (total); Asia, Australia (partial)
Annular; September 1, 2016	Parts of the Atlantic Ocean, Madagascar, central Africa, Indian Ocean (annular); continent of Africa (partial)
Annular; February 26, 2017	Parts of the Pacific and Atlantic Oceans, Argentina, Chile (annular); South America, Antarctica, Africa (partial)
Total; August 21, 2017	Parts of the United States, northern Pacific Ocean, Atlantic Ocean (total); North America, South America (partial)
Partial; February 15, 2018	Southern Cone, Antarctica
Partial; July 13, 2018	Southern Australia
Partial; August 11, 2018	Northeast Asia, northern Europe
Partial; January 6, 2019	Northern Pacific, northern Asia
Total; July 2, 2019	Southern Pacific Ocean, parts of Chile and Argentina (total); South America (partial)
Annular; December 26, 2019	India, Sumatra, Borneo, Saudi Arabia (annular); Asia, Australia (partial)

TYPE & DATE	WHERE VISIBLE
Annular; June 21, 2020	Southern Asia, central Africa, China (annular); rest of Africa, southern Europe, Asia (partial)
Total; December 14, 2020	Southern Pacific Ocean, Chile, Argentina (total); South America, Antarctica (partial)
Annular; June 10, 2021	Northern Canada, Russia, Greenland (annular); Iceland, North America, Europe, Asia (partial)
Total; December 4, 2021	Antarctica (total); southern Africa, Antarctic Ocean (partial)
Partial; April 30, 2022	Southern Pacific Ocean, South America
Partial; October 25, 2022	Northern Africa, Europe, Middle East, Western Asia
Total and annular; April 20, 2023	Australia, Papua New Guinea, Indonesia (total/annular); southern Asia, East Indies, Australia, Philippines, New Zealand (partial)
Annular; October 14, 2023	Western United States, Colombia, Brazil, Central America (annular); the rest of the Americas (partial)
Total; April 8, 2024	Mexico, eastern Canada, central United States (total); rest of North and Central America (partial)
Annular; October 2, 2024	Chile, Argentina (annular); South America (partial)

APPENDIX B

TYPE & DATE	WHERE VISIBLE
Total; September 28, 2015	North America, Europe, eastern Pacific Ocean
Penumbral; March 23, 2016	Western North America
Penumbral; September 16, 2016	Europe, western Pacific Ocean
Penumbral; February 11, 2017	North America, Europe
Partial; August 7, 2017	Europe
Total; January 31, 2018	Western North America, Pacific Ocean
Total; July 27, 2018	Europe
Total; January 21, 2019	North America, Europe, central Pacific Ocean
Partial; July 16, 2019	Europe
Penumbral; January 10, 2020	Europe
Penumbral; June 5, 2020	Europe
Penumbral; July 5, 2020	North America, southern Europe

TYPE & DATE	WHERE VISIBLE
Penumbral; November 30, 2020	North America, Pacific Ocean
Total; May 26, 2021	North America, Pacific Ocean
Partial; November 19, 2021	North America, northern Europe, Pacific Ocean
Total; May 16, 2022	North America, Europe
Total; November 8, 2022	North America, Pacific Ocean
Penumbral; May 5, 2023	Southern Hemisphere only
Partial; October 28, 2034	Eastern North America, Europe
Penumbral; March 25, 2024	North America
Partial; September 18, 2024	North America, Europe
Total; March 14, 2025	North America, western Europe, Pacific Ocean

GLOSSARY

ALTITUDE A reference point in the horizontal coordinate system used along with **azimuth** for determining a celestial body's location. Also known as elevation, altitude is measured in degrees between 0 and 90 and is the relative position of the object in terms of its height above the horizon. Any object below 0 degrees is below the horizon and not visible.

APPARENT MAGNITUDE The brightness of an object as viewed from Earth. Distant luminous objects may have lower apparent magnitudes than nearby dim objects; a jet plane, for instance, has a greater apparent magnitude than the Andromeda Galaxy.

ARCHAEOASTRONOMY Refers to the study of astronomy during pre-modern eras, often using ancient texts and artifacts, and the relationship of astronomy to culture. The Great Pyramids and Stonehenge are examples of artifacts that illustrate the astronomical knowledge of the era.

ARCMINUTE Defined as one-sixtieth of a degree. There are 60 arcminutes in a degree, and they are denoted with a single prime symbol ('). The Moon is about 30 arcminutes across, or one-half degree.

ARCSECOND Defined as one-sixtieth of an arcminute. There are 60 arcseconds in an arcminute, and they are denoted with a double prime (").

AZIMUTH One of two coordinates used in the horizontal coordinate system for determining a celestial body's location, along with **altitude**. The azimuth is the angle from north, measured in degrees from 0 to 360 and expressing how far east, from due north, the object is located.

BLACK HOLE A former star that had so much interior mass it collapsed into a point of mass so gravitationally powerful that light could not escape. Black holes are not directly visible, but they are usually surrounded by orbiting objects that occasionally fall into the black hole and briefly flash bright.

CATADIOPTRIC TELESCOPE A kind of hybrid telescope that combines elements of both refractors and reflectors.

CEPHEID VARIABLE A type of variable star that varies in brightness with a regular frequency. Cepheids are notable because the regularity of their cycles allows for careful measurements of distance.

CLUSTER One of several types of bright groupings of stars in the sky so named because they appear as a tightly knit group of bright objects. Some are the result of coincidence, though, such as open clusters that result when stars mutually form from the same nebula. Globular clusters are groups of stars orbiting a larger parent galaxy. There are other types that are much larger bodies of objects—specifically, galactic clusters, groups of galaxies orbiting each other.

COMPUTERIZED TELESCOPE A telescope with digitized elements—either coordinate systems, a camera, or controls. Many computerized telescopes require a power source, which can limit mobility.

DECLINATION Often abbreviated D. One of two measurements used in the equatorial coordinate system for determining a celestial body's location. Declination indicates how far up or down the object is in the sky, and is measured in degrees from -90 to 90 (see also **right ascension**).

DUST LANES In the context of nebulae and galaxies, dust lanes are dark spots that appear to obscure part of the brighter regions, much like a coating of dust on a television screen. Dust lanes often reveal information about the composition and origin of celestial bodies.

ECLIPTIC The path through the sky across which all the planets seem to pass as they orbit the Sun. This appears so because all planets orbit along the same plane; that is, if you were to look at them edge-on, they would line up. This hints at a common origin for all the planets in the solar system, presumably all formed from the same nebular material spinning at about the same speed. The 12 constellations that appear in the background of the ecliptic comprise the zodiac.

ELONGATION For planets closer to the Sun than to Earth, elongation is the angle formed by imagining a line between Sun, Earth, and the planet, with Earth in the middle. The point of greatest elongation occurs when the planet in question appears as far from the Sun as possible.

EQUATORIAL COORDINATE SYSTEM The most common astronomical coordinate system for finding objects in the night sky. It uses two measurements: **right ascension** and **declination**.

HELIOCENTRIC Means "Sun centered," as in the model of the solar system in which the Earth and other planets orbit the Sun. Galileo was one of the first to definitively prove Earth was part of a heliocentric solar system, rather than a geocentric (Earth-centered) one.

HEMISPHERE One of the two halves of the Earth, above and below the equator, split into Northern and Southern. The stars in the sky vary depending on which hemisphere you are in, as there are some stars never seen from the other. As such, astronomy is often discussed in reference to the two.

HORIZONTAL COORDINATE SYSTEM A less common but more robust system of astronomical coordinates. Relies on an **altitude** between 0 degrees and 90 degrees, and an **azimuth** between 0 degrees and 360 degrees that expresses how far east an object is.

LENS A molded object that converges light to a point. The human eye has a biological lens within, while telescopes typically have glass or plastic lenses.

LIGHT-MINUTE Measure of distance, specifically the distance an unimpeded ray of light travels in one minute. Since the speed of light is approximately 300,000 kilometers (186,000 miles) per second, a light-minute equals 17,987,548 kilometers (11,176,944 miles), or about one-tenth the distance between the Earth and Sun.

LIGHT POLLUTION The phenomenon of stray light, usually from cities or other human activity, reducing the ability to perceive astronomical objects. Light from Earth's surface is scattered through its atmosphere, lightening the sky's appearance from the ground. This is why at higher altitudes with less atmosphere above, and farther away from very bright cities like Las Vegas, the sky is often clearer. The Moon is one major source of nonterrestrial light pollution.

LIGHT-YEAR Deceptively, a light-year is a measurement of distance, not time. Since light in a vacuum travels around 300,000 kilometers (186,000 miles) per second, a light-year is the distance that light travels in one year in a vacuum. A light-year is equivalent to 9.5 trillion kilometers (5 trillion miles).

MESSIER OBJECT Charles Messier was a French comet hunter who created a catalogue of about 100 objects that were *not* comets but could potentially be confused with them. Most of these objects were actually nebulae and galaxies and are some of the most beautiful and accessible celestial bodies visible. Many famous celestial bodies have a Messier number and designation, which is often shortened to *M* followed by the number. For instance, Messier object 81 is shortened to M81 (see page 131). A Messier Marathon is an amateur astronomy challenge to view all 100-plus objects; for more information, visit messier.seds.org /xtra/marathon/marathon.html.

MIRA STAR (OR MIRA VARIABLE STAR) A type of variable star with a long period of many months and a reddish appearance. A catalogue and prediction of Mira variables is online here: www.britastro.org/vss /mira_predictions.htm.

NEBULA A cloud of gas and dust, often the remnant of an exploded star. Many nebulae are creating new stars as the gas slowly accumulates.

NEUTRON STAR A tiny, massive, fast-spinning object made entirely of neutrons that emits radio waves at high frequency. Neutron stars are the aftermath of the largest star explosions in the universe. Not visible with home telescopes.

NOVA Less powerful than a supernova, a nova occurs when a white dwarf with excess gas explodes violently. Many become nebulae afterward.

OCCULTATION The covering, or blocking, of one celestial body by another, frequently the Sun or Moon.

OPPOSITION The point at which a planet appears at its farthest point from Earth, due to its relative position in its orbit around the Sun.

PERIOD Length of time it takes a regularly revolving object to return to its starting point, making one full revolution. Earth's period is one year (the amount of time it takes to go around the Sun). Our solar system's period—the amount of time it takes to revolve around the central black hole of the Milky Way—is approximately 300 million years.

POLE STAR A star that appears directly above the rotational pole of Earth, such that it appears fixed while other stars in the sky move throughout the night. Pole stars vary as we move through the Milky Way. Currently the Northern Hemisphere has a pole star (Polaris; see page 47), while the Southern Hemisphere does not.

PROPER MOTION The apparent speed at which an object moves through the sky. Satellites, for instance, have very high proper motion, while most stars have very slow proper motions. As such, the constellations seen today are roughly the same as they were thousands of years ago. The star with the highest proper motion is called Barnard's Star.

RED DWARF A dim, small, and comparatively cool star. Few are visible with a home telescope.

RED GIANT A large, red star that has exhausted its primary fuel source (hydrogen), causing it to balloon to many times its original size and become much brighter. Many of the brightest nighttime stars are red giants.

REFLECTING TELESCOPE A type of telescope that uses curved mirrors to focus light.

REFRACTING TELESCOPE A type of telescope that uses curved lenses to focus light.

RETROGRADE A night sky object is said to be in retrograde when, after being observed night after night, it appears to briefly move in a direction opposite to that in which planets normally move. When Mercury is in retrograde, it appears to move west to east, rather than east to west.

REVOLUTION The process of orbiting around another object. Compare to **rotation.**

RIGHT ASCENSION Often abbreviated RA. One of two measurements used in the equatorial coordinate system for determining a celestial body's location. Right ascension determines how far the object is to the left or right and is measured in hours, as in 24 hours in a day (see also **declination**).

ROTATION The process of spinning on one's axis. Compare to **revolution.**

SOLAR FILTER A special fitting that screws into the end of your telescope and blocks out 99.9 percent of all light, allowing you to view the intricacies of the Sun without damaging your eyes.

SPLITTING The act of discerning a multiple-object system as its constituent parts, as opposed to one bright object. (See Sirius, page 53.)

STEREOSCOPIC Vision that involves both eyes rather than just one.

SUPERNOVA The brief and extremely powerful explosion of a star that has accumulated so dense an inner mass that it collapses in on itself, resulting in an implosion and then explosion. Supernovae are the most energy-intensive events in the universe, and even distant ones can be seen from Earth.

TOTALITY The moment during an eclipse when the occluding intermediate body completely covers the other body.

VARIABLE STARS Stars that oscillate in brightness with a predictable period.

WHITE DWARF Tiny but extremely bright stars created when a larger star's outer layers blow off in a nova or red giant explosion.

ZODIAC The 12 constellations symbolically important to astrology that appear in the background of the ecliptic.

RESOURCES

GENERAL AMATEUR ASTRONOMY RESOURCES

H. A. Rey's *The Stars:* the most useful resource for getting to know and identify the constellations of the night sky.

GENERAL ASTRONOMY AND ASTROPHYSICS

Carl Sagan's *Cosmos*—both the book and television series: tell the fascinating and rich story of our universe.

FOR YOUNG ADULTS AND CHILDREN

Astronomy Online: an amateur astronomy resource site: astronomyonline.org

EarthSky: a general resource with many constellation and star-finding pages: earthsky.org

Night sky playing cards: a useful way to learn constellations, if you play cards regularly. Several companies make them, including Jonathan Poppele.

Pocket sky atlas: *Sky & Telescope* (www.shopatsky.com) sells a book-size *Pocket Sky Atlas* with much more information than a star finder can hold.

Star finder: a dynamic, usually circular, device that allows you to rotate an image of the sky to correlate it with the season to figure out what is in the night sky on a given night during a given season. Star finders are necessary resources for any amateur astronomer. There are many different brands, including a kid-friendly one by Toysmith (www.toysmith.com) and an advanced one by Weems & Plath (www.weems-plath.com).

The Klutz Guide to the Galaxy: contains many resources for the young amateur astronomer.

The Old Farmer's Almanac: keeps a yearly updated record of meteor showers: www.almanac.com/content/meteor-showers-guide

MAGAZINES

The two best astronomy magazines are listed below. Both have frequently updated star charts, constellation finders, and many details and reviews of telescopes and other viewing equipment.

Astronomy: www.astronomy.com

Sky & Telescope: www.skyandtelescope.com

TELESCOPE INFORMATION

Robin Scagell's book *Stargazing with a Telescope:* a great resource for purchasing a telescope and understanding how they work.

ASTROPHOTOGRAPHY

Amateur Astrophotography, a magazine by amateurs: www.amateurastrophotography.net

Tips and techniques for astrophotography: www.astropix.com

FORUMS FOR AMATEUR ASTRONOMERS

Astroholic: www.astroholic.com

Astromart: www.astromart.com/forums

Cloudy Nights: www.cloudynights.com

SOUTHERN HEMISPHERE STARGAZING RESOURCES

Australian Amateur Astronomer Forum: www.iceinspace.com.au/forum

David Chandler makes a nice Southern Hemisphere night sky star finder: www.davidchandler.com

Amateur Association of Variable Star Observers: www.aavso.org

List of Mira stars, and predictions as to variability: www.britastro.org/vss/mira_predictions.htm

COMETS

List of bright Northern Hemisphere comets, updated regularly: www.aerith.net/comet/future-n.html

Real-time comet updates and comments on currently visible comets: www.ast.cam.ac.uk/~jds

STORES

Amazon.com: carries nearly every major brand of telescope and filter, including Celstron, Orion, Twinstar, and Barska.

OpticsPlanet: major international dealer and shipper for telescopes, binoculars, and filters. www.opticsplanet.com/astronomy-store.html

Planetariums: most carry stargazing accessories, including small telescopes and star finders. Often planetariums also hold viewing parties and events. Find a comprehensive list of planetariums in the United States at www.go-astronomy.com/planetariums.htm.

Starizona: one of the best local dealers and maker of telescopes, as well. Located in Tucson, Arizona, a hotbed for astronomy and observing, Starizona has a physical store as well as an online one, and holds viewing parties frequently. www.starizona.com

ECLIPSES

NASA keeps a comprehensive record of upcoming eclipses on its website (www.nasa.gov):

Lunar eclipses: eclipse.gsfc.nasa.gov/lunar.html

Solar eclipses: eclipse.gsfc.nasa.gov/solar.html

REFERENCES

Beatty, Kelly. "Where, When, and How to See Mercury." *Sky & Telescope*, October 30, 2014. Accessed May 3, 2015. www.skyandtelescope.com/astronomy-news/observing-news/spot-mercury-at-dawn-103020143

Bicknell, Peter. "The Witch Algaonice and Dark Lunar Eclipses in the Second and First Centuries BC," *Journal of the British Astronomical Association* 93, no. 4 (1983): 160–3. doi:1983JBAA . . . 93..160B

Buchen, Lizzie. "May 29, 1919: A Major Eclipse, Relatively Speaking." *Wired.* May 29, 2009. www.wired.com/2009/05/dayintech_0529

Burnham, R. *Burnham's Celestial Handbook, Volume Three: An Observer's Guide to the Universe Beyond the Solar System* (p. 1,955). Newburyport, MA: Dover Publications, 1977.

Carroll, Bradley W., and Dale A. Ostlie. *An Introduction to Modern Astrophysics.* 2nd ed. San Francisco, CA: Pearson Addison-Wesley, 2007.

Consolmagno, Guy, and Dan M. Davis. *Turn Left at Orion.* Cambridge, UK: Cambridge University Press, 2000.

Corbett, Bill. *A Simple Guide to Telescopes, Spotting Scopes, and Binoculars.* New York, NY: Watson-Guptill Publications, 2003.

Dickinson, Terence. *NightWatch: A Practical Guide to Viewing the Universe.* 3rd ed. Willowdale, Ontario: Firefly Books, 1998.

Meara, Stephen James. *Deep-Sky Companions: The Messier Objects.* 2nd ed. New York, NY: Cambridge University Press, 2014.

Messier, Charles. *The Messier Catalog.* Tucson, AZ: SEDS, 1994.

Motz, Lloyd, and Carol Nathanson. *The Constellations.* New York, NY: Doubleday, 1988.

National Aeronautics and Space Administration. "NASA Lunar Eclipse Page." Accessed May 4, 2015. eclipse.gsfc.nasa.gov/lunar.html

National Aeronautics and Space Administration. "NASA Solar Eclipse Page." Accessed May 4, 2015. eclipse.gsfc.nasa.gov/solar.html

Raymo, Chet. *365 Starry Nights*. Englewood Cliffs, NJ: Prentice-Hall, 1982.

Rey, H. A. *The Stars: A New Way to See Them*. New York, NY: Houghton Mifflin Harcourt, 2008.

Sagan, Carl. *Cosmos*. New York, NY: Random House, 1980.

Scagell, Robin. *Stargazing with a Telescope*. New York, NY: Cambridge University Press, 2000.

Schove, D. Justin. *Chronology of Eclipses and Comets: AD 1–1000*. Dover, NH: Boydell, 1984.

Sigismondi, C., D. Hoffleit, and R. Coccioli. "Long-Term Behavior of Mira Ceti Maxima." *The Journal of the American Association of Variable Star Observers* 30, no. 1 (2001): 31–3. doi:2001JAVSO..30...31S

Society for Popular Astronomy. "Guide to Betelgeuse (Alpha Orionis)." Accessed May 2, 2015. www.popastro.com/variablestar/reference/guide.php?title_pag=Guidepercent 20topercent 20Betelgeusepercent 20percent 28Alphapercent 20Orionispercent 29

Stoyan, Ronald. *Atlas of the Messier Objects: Highlights of the Deep Sky*. Cambridge, UK: Cambridge University Press, 2008.

Suomi, V. E., S. S. Limaye, and D. R. Johnson. "High Winds of Neptune: A Possible Mechanism." *Science* 251, no. 4996 (1991): 929–32. doi:10.1126/science.251.4996.929

Whitman, Alan. "Digging Deep in M33." *Sky & Telescope*, December 2004, 92–96.

INDEX

CPSIA information can be obtained
at www.ICGtesting.com
Printed in the USA
JSHW021906091021
19373JS00001B/3